CONDUCT OF OPERATIONS AND OPERATIONAL DISCIPLINE

This book is one in a series of process safety guideline and concept books published by the Center for Chemical Process Safety (CCPS). Please go to www.wiley.com/go/ccps for a full list of titles in this series.

CONDUCT OF OPERATIONS AND OPERATIONAL DISCIPLINE

For Improving Process Safety in Industry

Center for Chemical Process Safety
New York, New York

A Joint Publication of the Center for Chemical Process Safety of the American Institute of Chemical Engineers and John Wiley & Sons, Inc.

Published by John Wiley & Sons, Inc., Hoboken, New Jersey.
Published simultaneously in Canada.

Library of Congress Cataloging-in-Publication Data:

Conduct of operations and operational discipline : for improving process safety in industry.
p. cm.
"Center for Chemical Process Safety"—T.p.
Includes bibliographical references and index.
ISBN 978-0-470-76771-9 (hardback)
1. Industrial safety. I. American Institute of Chemical Engineers. Center for Chemical Process Safety.
T55.G788 2011
660'.2804—dc22 2010047225

Printed in the United States of America.

oBook: 978-1-118-029206
ePDF: 978-1-118-029183
ePub: 978-1-118-029190

SKY10089217_102824

CONTENTS

7 IMPLEMENTING AND MAINTAINING EFFECTIVE COO/OD SYSTEMS 167

LIST OF TABLES

LIST OF FIGURES

ONLINE MATERIALS ACCOMPANYING THIS BOOK

Chapter	Associated Online Material Accompanying This Book
Chapter 1	• Diagrams from Chapter 1 in a Microsoft® PowerPoint® presentation • COO/OD System Inputs and Outputs for RBPS Elements (includes additional RBPS elements not addressed in Table 1.7)
Chapter 2	None
Chapter 3	None
Chapter 4	• Additional Examples of Error-Likely Situations (includes additional examples similar to those in Table 4.3 of the book) • The Three Dimensions of Consequences (material that supplements Section 4.8 of the book)
Chapter 5	None
Chapter 6	None
Chapter 7	• Additional Metrics (list that supplements the metrics listed in Section 7.4.1 of the book) • COO Survey (courtesy of Concord Associates, Inc.)

To access this online material, go to www.aiche.org/ccps/publications/COOD.aspx

Enter the password: COOD2010

ACRONYMS AND ABBREVIATIONS

ABC	antecedent-behavior-consequence
ACC	American Chemistry Council
AIChE	American Institute of Chemical Engineers
API	American Petroleum Institute
ATM	automated teller machine
BB	behavior-based
CCPS	Center for Chemical Process Safety
ClO_2	chlorine dioxide
CO	commanding officer
COO	conduct of operations
CRM	crew resource management
CSB	U.S. Chemical Safety and Hazard Investigation Board
DOE	U.S. Department of Energy
Dow	Dow Chemical Company
DSEAR	Dangerous Substances and Explosive Atmospheres Regulations
DuPont	E. I. du Pont de Nemours and Company
EH&S	environmental, health, and safety
EPA	U.S. Environmental Protection Agency
FDA	U.S. Food and Drug Administration
GM	General Motors Company
HF	hydrofluoric acid

HPCL	Hindustan Petroleum Corporation Ltd.
HPT	human performance technology
INPO	Institute of Nuclear Power Operations
ISO	International Organization for Standardization
ITPM	inspection, test, and preventive maintenance
JSA	job safety analysis
MOC	management of change
N-D-C	negative, delayed, certain
N-D-U	negative, delayed, uncertain
N-I-C	negative, immediate, certain
N-I-U	negative, immediate, uncertain
NASA	U.S. National Aeronautics and Space Administration
NRC	U.S. Nuclear Regulatory Commission
NUMMI	New United Motor Manufacturing, Inc.
Occidental	Occidental Petroleum (Caledonia) Ltd.
OD	operational discipline
OOD	officer on deck
OSHA	U.S. Occupational Safety and Health Administration
P-D-C	positive, delayed, certain
P-D-U	positive, delayed, uncertain
P-I-C	positive, immediate, certain
P-I-U	positive, immediate, uncertain
PD	periscope depth
PDCA	Plan-Do-Check-Adjust
PHA	process hazard analysis
PPE	personal protective equipment
PSM	process safety management
PSV	pressure safety valve
R&D	research and development
RBPS	risk-based process safety
RMP	risk management program
RP	Recommended Practice

SMART	Specific, Measurable, Attainable, Relevant, Time-specific
SRK	skill, rule, knowledge
STAR	Stop, Think, Act, and Review
SWP	safe work practice
Toyota	Toyota Motor Corporation
U.K.	United Kingdom
VCM	vinyl chloride monomer

GLOSSARY

Antecedent-behavior-consequence (ABC) analysis: A human performance analysis tool that examines how human behavior is influenced by previous experiences with similar situations and expectations of reward or punishment.

Balanced scorecard: A management system that provides feedback on both internal business processes and external outcomes to continuously improve strategic performance and results.

Behavior-based safety program: A program designed to provide frequent feedback to personnel regarding their safety behaviors in the workplace.

Conduct of operations (COO): The embodiment of an organization's values and principles in management systems that are developed, implemented, and maintained to (1) structure operational tasks in a manner consistent with the organization's risk tolerance, (2) ensure that every task is performed deliberately and correctly, and (3) minimize variations in performance.

- COO is the management systems aspect of COO/operational discipline (OD).
- COO sets up organizational methods and systems that will be used to influence individual behavior and improve process safety.
- COO activities result in specifying how tasks (operational, maintenance, engineering, etc.) should be performed.
- A good COO system visibly demonstrates the organization's commitment to process safety.

Consequence: Within the context of human performance, the direct and indirect results of an action.

Deviation: A variation in data, process variables, or human action that is large enough to exceed established design limits, safe operating limits, or standard operating procedures.

Discipline: Within the context of OD, discipline refers to (1) an orderly or prescribed conduct or pattern of behavior and (2) a rule or system of rules governing conduct or activity. The word "discipline," as used in OD, does NOT refer to punishment.

Error-likely situation: A work situation in which the performance-shaping factors are not compatible with the capabilities, limitations, or needs of the operator. This situation is likely to prevent the operator from correctly performing the task.

Error-proofing: Use of process or design features to prevent the occurrence, further processing, or acceptance of nonconforming actions or products. Also known as "mistake-proofing."

Fixed facility: A portion of or a complete plant, unit, site, complex or any combination thereof that is generally not moveable. In contrast, mobile facilities, such as ships (e.g., transport vessels, floating platform storage and offloading vessels, drilling platforms), trucks, and trains, are designed to be movable.

Front-line personnel: The personnel who perform tasks that produce the output of the work group. Front-line personnel include operations and maintenance personnel, engineers, chemists, accountants, shipping clerks, etc.

Human error:

1. Any human action (or lack thereof) that exceeds some limit of acceptability (i.e., an out-of-tolerance action) where the limits of human performance are defined by the system. Includes actions by designers, operators, or managers that may contribute to or result in accidents.
2. Mistakes by people, such as designers, engineers, operators, maintenance personnel, or managers, that may contribute to or result in hazardous events and incidents.

Human factors:

1. A discipline concerned with designing machines, operations, and work environments so that they match human capabilities, limitations, and needs. Includes any technical work (engineering, procedure writing, worker training, worker selection, etc.) related to the human factor in operator-machine systems.
2. Selecting materials or equipment that can better tolerate human error in handling; making a process or piece of equipment easier to understand, easier to function as intended, or more difficult to function improperly; ergonomics.

Human performance technology: A systematic approach to improving productivity and competence that uses a set of methods and procedures to realize opportunities related to the performance of people.

Incident: An unplanned event or series of events and circumstances that may result in an undesirable consequence, such as injury to personnel, damage to property, adverse environmental impact, or interruption of process operations.

Knowledge-based behavior: Performance that requires personnel to consciously select and execute actions.

Lagging indicators: Outcome-oriented metrics, such as incident rates, downtime, quality defects, or other measures of past performance.

Leading indicators: Process-oriented metrics, such as the degree of implementation of or conformance with policies and procedures that support a management system.

Management system:

1. An administrative system that governs essential business activities.
2. A formally established set of activities designed to produce specific results in a consistent manner on a sustainable basis.
3. A program or activity involving the application of management principles and analytical techniques to ensure that the core attributes of each protection layer are met.

Mental models: An individual's or group's simplified representation of a process or system that explains the relationship between its various inputs, internal processes, and outputs.

Mitigation safeguards: A safeguard that is designed to reduce the severity of a loss event. Mitigation safeguards can be divided into detection safeguards and correction safeguards.

Operational discipline (OD): The performance of all tasks correctly every time.

- OD is the execution of the COO system by individuals within the organization.
- OD refers to the day-to-day activities carried out by all personnel.
- Individuals demonstrate their commitment to process safety through OD.
- Good OD results in performing the task the right way every time.
- Individuals recognize unanticipated situations, keep (or put) the process in a safe configuration, and seek involvement of wider expertise to ensure personal and process safety.

Organizational culture: The common set of values, behaviors, and norms at all levels in a facility or in the wider organization that affect the operation of the facility.

Plan-Do-Check-Adjust (PDCA) approach: A four-step process for quality improvement. In the first step (Plan), a way to bring about improvement is developed. In the second step (Do), the plan is carried out. In the third step (Check), what was predicted is compared to what was observed in the previous step. In the last step (Adjust), plans are revised to eliminate performance gaps. The PDCA cycle is sometimes referred to as (1) the Shewhart cycle because Walter A. Shewhart discussed the concept in his book entitled Statistical Method from the Viewpoint of Quality Control or (2) the Deming cycle because W. Edwards Deming introduced the concept in Japan; the Japanese subsequently called it the Deming cycle. It is also called the Plan-Do-Study-Act (PDSA) cycle.

Preventive safeguards: A safeguard that forestalls the occurrence of a particular loss event, given that an initiating cause has occurred; i.e., a safeguard that intervenes before an initiating cause can produce a loss event.

Process life cycle: The stages that a physical process or a management system goes through as it proceeds from birth to death. These stages include conception, design, deployment, acquisition, operation, maintenance, decommissioning, and disposal.

Process safety culture: The common set of values, behaviors, and norms at all levels in a facility or in the wider organization that affect process safety.

Repeat-back: A method of communication that requires the receiver to repeat the message back to the sender to validate that the appropriate message was received.

Risk-based process safety (RBPS): The Center for Chemical Process Safety's process safety management system approach that uses risk-based strategies and implementation tactics that are commensurate with the risk-based need for process safety activities, availability of resources, and existing process safety culture to design, correct, and improve process safety management activities.

Risk tolerance: The maximum level of risk of a particular technical process or activity that an individual or organization accepts to acquire the benefits of the process or activity.

Rule-based behavior: Behavior in which a person follows remembered or written rules. Examples might be the use of a written checklist to calibrate an instrument or the use of a maintenance manual to repair a pump.

Safeguard: Any device, system, or action that would likely interrupt the chain of events between an initiating cause and a specific loss event.

Skill-based behavior: The performance of routine actions governed by stored patterns of behavior. Examples might be the use of a hand tool by an experienced mechanic or the initiation of an emergency procedure by a trained and experienced operator.

SMART: Specific, Measurable, Attainable, Relevant, Time-specific. Other potential meanings: S – significant, stretching; M – meaningful, motivational; A – agreed upon, acceptable, action-oriented; R – realistic, reasonable, rewarding, results-oriented; T – timely, tangible, trackable, time-bound.

Thoughtful compliance: Performing tasks in compliance with all rules and requirements, but seeking the involvement of wider expertise when existing rules and requirements appear to be in conflict with process safety goals.

Variation: A change in data, process parameter, or human behavior. Within prescribed limits, changes in data, process parameters, and human behavior are anticipated and acceptable. Variation outside established limits is called deviation.

World-class manufacturing: A position of international manufacturing excellence, achieved by developing a culture based on factors such as continuous improvement, COO/OD, problem prevention, zero defect tolerance, customer-driven just-in-time production, and total quality management.

ACKNOWLEDGMENTS

The American Institute of Chemical Engineers (AIChE) and the Center for Chemical Process Safety (CCPS) express their gratitude to all of the members of the Conduct of Operations/Operational Discipline Subcommittee and their CCPS member companies for their generous efforts and technical contributions in the preparation of this *Concept* book.

The chairman of the Subcommittee was James Klein from DuPont. Greg Keeports was the CCPS staff liaison. The Subcommittee also included the following people who participated in the writing of this book:

Guy Arnaud	TOTAL TS
John Herber	3M (retired)
Mark Leigh	ConocoPhillips
Robin Pitblado	DNV

The following people participated in the original Subcommittee that structured this *Concept* book:

Rob DiValerio	BP
Niamh Donohoe	Intel
John Haesle	Celanese
Lou Higgins	Rhodia
Karen Paulk	ConocoPhillips
Fran Schultz	SABIC Innovative Plastics
Greg Schultz	Dow Chemical
Gary Stubblefield	Baker Risk

CCPS wishes especially to acknowledge the contributions of the principal authors from ABSG Consulting Inc. (ABS Consulting):

Bill Bradshaw
Don Lorenzo
Lee Vanden Heuvel, *Project Manager*

The authors wish to thank the following ABS Consulting personnel for their technical contributions and reviews: James Liming provided technical review of the book. Leslie Adair edited the manuscript. Paul Olsen created many of the graphics. Finally, Susan Hagemeyer prepared the final manuscript for publication.

Before publication, all CCPS books are subjected to a thorough peer-review process. CCPS also gratefully acknowledges the thoughtful comments and suggestions of the following peer reviewers. Their work enhanced the accuracy, clarity, and usefulness of this *Concept* book.

Mark Begg	Air Products
Mike Broadribb	Baker Risk
Lalaine Byrd	Intel
Jack Chosnek	KnowledgeOne
Lloyd Cowlam	ConocoPhillips
Art Dowell	Dow/Rohm & Haas (retired)
Rick Ewan	Stonehill Consulting, LLC
Jeffrey Fox	Dow Corning
Pete Lodal	Eastman Chemicals
M Fazaly M Ali	Petronas
Sam Mannan	Mary Kay O'Connor Process Safety Center
Jack McCavit	JLM Consulting
Mickey Norsworthy	Process Improvement Institute
Jack Philley	Baker Hughes/Baker Petrolite
Rich Purgason	LyondellBasell
Ronald Rhodes	TOTAL Petrochemicals
David Thaman	PPG Industries
Lee Valentine	BP
Bruce Vaughen	Cabot
Terry Welch	BP

PREFACE

The American Institute of Chemical Engineers (AIChE) has been closely involved with process safety and loss control issues in the chemical and allied industries for more than four decades. Through its strong ties with process designers, constructors, operators, safety professionals, and members of academia, the AIChE has enhanced communications and fostered continuous improvement of the industry's high safety standards. AIChE publications and symposia have become information resources for those devoted to process safety and environmental protection.

The AIChE created the Center for Chemical Process Safety (CCPS) in 1985 after the chemical disasters in Mexico City, Mexico, and Bhopal, India. CCPS is chartered with developing and disseminating technical information for use in the prevention of major chemical accidents. The center is supported by more than 125 industry sponsors who provide the necessary funding and professional guidance to its technical committees. The major product of CCPS activities has been a series of guidelines and essential practices to assist those implementing various elements of a process safety and risk management system.

This book is part of the *Concept* series of books that are focused on specific topics and are intended to complement the longer, more comprehensive *Guidelines* series of books.

Conduct of operations (COO) was first proposed by CCPS in 2007 as a process safety element in the *Guidelines for Risk Based Process Safety*, which updated the original CCPS guidance to reflect 15 years of process safety management (PSM) implementation experience, best practices from relevant industries, and global regulatory requirements. COO was added because other elements of process safety are only effective if there is system to ensure reliable, consistent, and correct execution of the policies, procedures, and practices that make up the facility's risk management system.

COO does not focus on basic operations and maintenance elements, such as procedures, training, safe work practices, asset integrity, management of change, and pre-startup safety review. Rather, it is a management system to help ensure the effectiveness of these and other PSM systems.

For this book, the element was split into COO and operational discipline (OD). COO encompasses the ongoing management system aspects while OD is the deliberate and structured execution of the COO system by individuals at every level of the organization, starting at the top. This book provides specific guidance on how an effective COO/OD system can be established and implemented. However, COO/OD is not a "quick fix" solution – success requires an enduring commitment from the organization's leadership team. If you are just getting started with COO/OD, you should find all of the chapters helpful. If your organization's management is already supportive of COO/OD and you are just looking for specific actions to implement, focus on Chapters 5, 6, and 7.

EXECUTIVE SUMMARY

Process safety practices and formal safety management systems have been in place in some companies for more than 100 years. Process safety management (PSM) is widely credited for reductions in major accident risk and for improved chemical industry performance. Nevertheless, many organizations are still challenged with effectively implementing the management systems they have developed. This *Concept* book is intended to improve the execution of PSM elements in the process and allied industries.

The purpose of this book is to help organizations design and implement **conduct of operations** (COO) and **operational discipline** (OD) systems. This book provides ideas and methods on how to (1) design and implement COO and OD systems, (2) correct deficient COO and OD systems, or (3) improve existing COO and OD systems.

> COO addresses management systems. OD addresses the execution of the COO and other management systems.

In general, COO encompasses the ongoing management systems that are developed to encourage performance of all tasks in a consistent, appropriate manner. OD is the deliberate and structured execution of the COO and other organizational management systems by personnel throughout the organization. Formal definitions of COO and OD can be found in Section 1.4.

Figure S.1 shows a process safety pyramid or triangle, where the minor, serious, and catastrophic injuries normally found progressing up to the top of a personal safety triangle have been replaced with appropriate process safety issues, consistent with the process safety focus of this book. Eliminating the issues at the base of the triangle should result in a reduction in process safety incidents. COO/OD activities are typically focused on the bottom portion of the triangle with the goal of reducing the number of issues that occur at higher levels of the triangle.

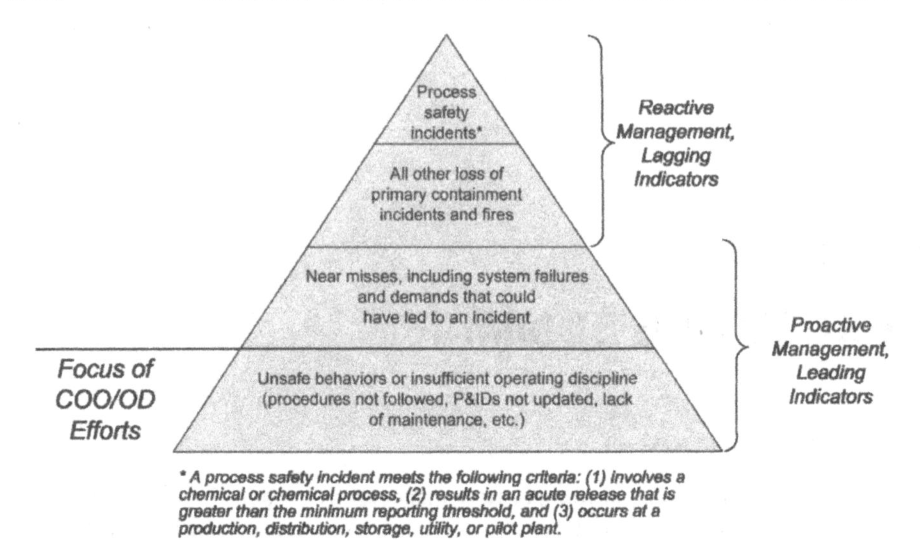

FIGURE S.1. Typical Process Safety Pyramid

Key attributes of COO systems include:

People

- Clear Authority/Accountability
- Communications
- Logs and Records
- Training, Skill Maintenance, and Individual Competence
- Compliance with Policies and Procedures
- Safe and Productive Work Environments
- Aids to Operation – the Visible Plant
- Intolerance of Deviations
- Task Verification
- Supervision/Support
- Assigning Qualified Workers
- Access Control
- Routines
- Worker Fatigue/Fitness for Duty

Process

- Process Capability
- Safe Operating Limits
- Limiting Conditions for Operation

Plant

- Asset Ownership/Control of Equipment
- Equipment Monitoring
- Condition Verification
- Management of Subtle Changes
- Control of Maintenance Work
- Maintaining the Capability of Safety Systems
- Controlling Intentional Bypasses and Impairments

Key attributes of OD systems include:

Organizational

- Leadership
- Team Building and Employee Involvement
- Compliance with Procedures and Standards
- Housekeeping

Individual

- Knowledge
- Commitment
- Awareness
- Attention to Detail

Figure S.2 illustrates the basic process used to implement a COO/OD system. The process can be entered from two conditions. The entry point at the top of the diagram is appropriate for a new COO/OD system. The second entry point, at the bottom of the diagram, is better suited to efforts to improve an existing COO/OD system. The first step for a new system is to establish (or revise) the goals and management leadership to make the system successful. Next, the COO/OD system is developed/revised and implemented. As the COO/OD system is implemented, its performance is measured. Based on the performance data, revisions are made to the COO/OD system. This cycle then continues as the system is monitored and improved over time.

ORGANIZATION OF THIS BOOK

Chapter 1 of the book provides definitions of conduct of operations and operational discipline, along with guidance on determining whether an improved COO/OD program is required within the organization. Chapter 2 outlines the benefits of implementing a COO/OD system. Chapter 3 describes the important role that management leadership has in successful implementation of the system. Chapter 4 describes human factors issues that are important in either setting up the system or in identifying solutions to performance problems. Chapter 5 describes key attributes of a COO system, and Chapter 6 describes key attributes of an OD system. Chapter 7 completes the COO/OD model by describing how to monitor its performance and continuously improve it.

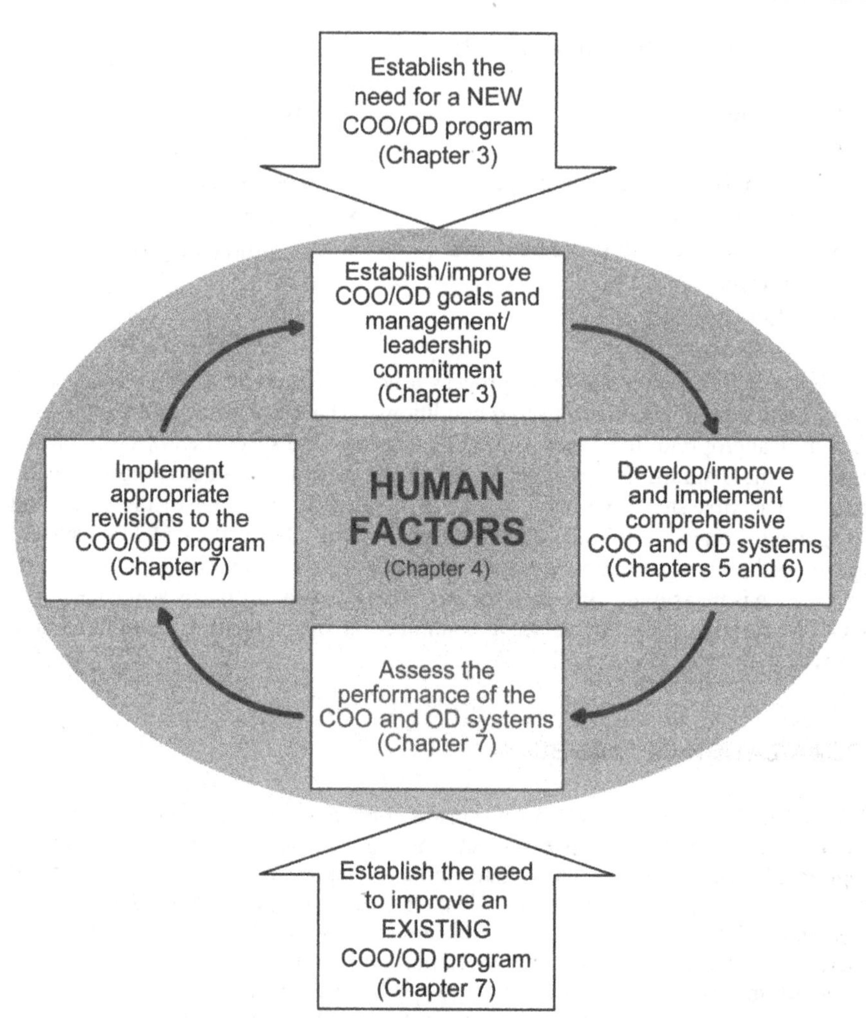

FIGURE S.2. COO/OD Improvement and Implementation Cycle

1
WHAT IS COO/OD AND HOW CAN I TELL IF I NEED IT?

1.1 INTRODUCTION

This book describes the concepts of conduct of operations (COO) and operational discipline (OD), the attributes of effective COO/OD systems, and the steps an organization might take to implement or improve its COO/OD systems. This chapter should be read by everyone using this book to familiarize themselves with the principles of COO/OD. It will explain the basic COO/OD concepts and help you decide whether your current COO/OD system activities need improvement. It will also define important terms used throughout the book and the relationship between COO/OD and other management systems.

In general, COO encompasses the ongoing management systems[1] that are developed to encourage performance of all tasks in a consistent, appropriate manner. OD is the deliberate and structured execution of the COO and other organizational management systems by personnel throughout the organization. Formal definitions of COO and OD can be found in Section 1.4.

> COO addresses management systems. OD addresses the execution of the COO and other management systems.

1.2 PURPOSE OF THIS BOOK

This *Concept* book is intended to explain the key attributes of COO/OD and to provide specific guidance on how an organization can implement effective systems.

The purpose of this book is to help organizations design and implement COO and OD systems. This book provides ideas and methods on how to (1) design and implement COO and OD systems, (2) correct deficient COO and OD systems, or (3) improve existing COO and OD systems.

1.3 FOCUS AND INTENDED AUDIENCE

The primary focus of this book is on improving process safety management within the process and allied industries. However, the concepts and activities described in this book should be applicable to a broad spectrum of facilities in many industries.

---

[1] Organizations typically use the term "program" or "system" to describe their approach to COO/OD. The term "system is used in this book. One term that should not be used is COO/OD "project"; COO/OD is not a project with a discrete end date, but an ongoing process.

Its intended audience is everyone – from upper management to front-line workers – who will be involved in designing, implementing, maintaining, and improving COO/OD systems. Section 1.5 discusses how the intended audience might use this book.

> **PSM USAGE**
>
> The terms "process safety management" and "PSM," as used throughout this book, refer to the systems used to manage process safety within an organization. They do **NOT** refer to a specific regulation (such as 29 CFR 1910.119 in the United States).

Implementing an effective COO/OD system inevitably produces positive changes in an organization's culture; however, changing the overall culture of an organization is a broader topic than the COO/OD systems addressed herein. Likewise, the broad application of COO/OD principles will likely produce occupational safety, environmental, reliability, quality, and many other benefits. However, this book focuses on the process safety aspects of COO/OD. The examples used throughout the book and the work activities described emphasize process safety issues.

> **PROCESS SAFETY FOCUS**
>
> This book focuses on improving process safety performance, which may also bring occupational safety benefits.

> **BP Texas City – An Example of COO/OD Failings**
>
> On March 23, 2005, an explosion occurred in the Isomerization Unit (ISOM) at the BP refinery in Texas City, Texas, during a startup after a turnaround (Ref. 1.1). The incident resulted in 15 fatalities, more than 170 people injured, and major damage to the ISOM and adjacent process units.
>
> The vapor cloud explosion occurred after liquid hydrocarbons were ejected from the stack of the blowdown drum serving the ISOM raffinate splitter column, which had been overfilled.
>
> COO/OD-related issues associated with this incident include the following:
>
> - An operational check of the independent high level alarm in the raffinate splitter tower was not performed prior to startup, even though it was required by procedures.
> - The operators did not respond to the high level alarm in the splitter (it was on throughout the incident).
> - The level indication available to the operators was useless during most of the startup because they deliberately maintained the level above the indicated range of the level instruments.
> - When the Day Shift Supervisor arrived at about 7:15 a.m., no job safety review or walkthrough of the procedures to be used that day was performed as required by procedures.

- The board operator printed off the wrong startup procedure (although this was not a significant factor because he never referred to it).
- The splitter bottoms were heated at 75°F per hour despite the procedural limit of 50°F per hour.
- The Day Shift Supervisor left the plant during the startup about 3½ hours prior to the explosion. No replacement was provided during this period.
- The operating procedures were certified as current, although they did not include changes to relief valve settings made prior to the most recent recertification.
- Outside operators did not report significant deviations of operating parameters (such as rising pressure on the splitter bottoms pumps) to the control room.
- Deficiencies first identified in 2003 and 2004 still existed in training programs for ISOM operators.

Other notable examples of incidents with significant COO/OD issues include the following:

- Three Mile Island nuclear plant incident, March 28, 1979 (Ref. 1.2)
- Union Carbide methyl isocyanate release, Bhopal, India, December 3, 1984 (Ref. 1.3)
- Chernobyl nuclear plant explosion, April 26, 1986 (Ref. 1.4)
- Piper Alpha oil production platform fire, July 6, 1988 (Ref. 1.5)
- *Exxon Valdez* oil tanker spill on Bligh Reef near Valdez, Alaska, March 24, 1989 (Ref. 1.6)
- Sinking of the Petrobras P-36 oil production platform in the Roncador Field, May 15, 2001 (Ref. 1.7)

In all of these incidents, the information needed to safely operate the facility was present in the procedures and practices of the facility or known by facility personnel. Yet, in every case, well-intentioned, well-trained workers committed grievous errors. Why didn't the facility personnel perform the work appropriately? One contributor to these incidents was a lack of an effective COO/OD system.

Consider an acid leak that developed unnoticed as a result of poor housekeeping. This book will focus on the process hazards associated with the acid leak, not on the company's culture of using only a proven technology requiring acid instead of an inherently safer, but unproven, acid-free alternative. If the worker was injured as a result of not wearing the proper personal protective equipment (PPE) at the time of the acid leak, this book will focus on the consequences of not being able to isolate the release quickly, not on the injury resulting from the operator being splashed with acid. But, as noted above, preventing the acid leak and routinely wearing the proper PPE would not only have process safety benefits, but also occupational safety benefits.

NEW ELEMENT OF RISK-BASED PROCESS SAFETY

In its 2007 *Guidelines for Risk Based Process Safety* (Ref. 1.8), the Center for Chemical Process Safety (CCPS) identified COO as an essential element of a comprehensive risk-based process safety (RBPS) management system. Incorporation of COO into the RBPS guidelines was based on a long history of formalized operations concepts at many companies. For this book, the element was split into COO and OD (see Chapter 2 for a more detailed history of COO/OD systems). The RBPS guidelines identified twenty RBPS elements and organized them into four pillars of process safety. The COO/OD element is included in the Managing Risk pillar. Chapter 17 of the RBPS guidelines outlines the key principles and essential features of the COO/OD element, and it lists more than fifty possible work activities related to the element (with associated implementation options), examples of ways to improve the effectiveness of the element, metrics, and management review activities related to the element.

The COO/OD system applies to all personnel in the organization, including direct-hire employees, contractors, third-party personnel, and part-time employees. All personnel must be included in a successful COO/OD system.

A fully implemented COO/OD system touches every level of an organization, from the boardroom to the shop floor. For example, the manner in which a Vice-President of Operations handles weekly management meetings and addresses specific process safety topics falls within the COO/OD system. Table 1.1 lists some examples of how the COO/OD system applies to management personnel.

Thus, this book is initially directed toward an organization's leadership team. The team must decide that the long-term benefits of COO/OD, described in Chapter 2, are worth the initial and ongoing investment. The book then describes COO/OD systems in detail, which enables upper management to estimate the costs and benefits of such systems so that they can make an informed decision on how to proceed. The book also helps management understand that it must make a visible ongoing commitment if the system is to succeed.

Once the organization decides to implement COO/OD, overall responsibility for implementation and maintenance of this system rests with the facility manager[2], although its concepts can also be applied at the corporate level. This book will help facility managers identify systems that they should implement as part of a comprehensive COO/OD system. The bulk of the book is intended for those managers and specialists who will be developing, implementing, and maintaining the COO/OD system. This book describes typical features of a COO/OD system so that the responsible parties can perform a gap analysis of their existing systems and then improve their systems or use the model programs as a starting point for developing their own (see Chapter 7). This book will help site operations leaders

[2] The facility manager is the individual who has overall accountability and responsibility for the safe and efficient operation of an asset. A variety of terms may be used at different types of facilities. For example, at a fixed production facility this person may have the title of Plant or Site Manager. For an offshore oil platform, this individual may be referred to as the Offshore Installation Manager.

and area managers define the framework of controls necessary to ensure that tasks for which they are responsible are performed reliably.

TABLE 1.1. Examples of Management Operational Discipline Resulting from a COO System

- Exploring process safety management performance and efficiency issues in a positive way
- Requiring the collection of key performance indicators for process safety and regularly reviewing them
- Setting process safety performance expectations and providing the resources to achieve them
- Looking for management system failures as root causes for incidents
- Consistently identifying and correcting substandard actions or conditions during field walkthroughs
- Completing management reviews and approvals related to work activities in a timely manner
- Communicating a meeting's purpose and agenda reasonably in advance and conducting meetings efficiently
- Treating peers and subordinates in a respectful manner
- Documenting the results of meetings and transmitting the minutes within a reasonable time
- Holding everyone (including themselves) accountable for commitments and ensuring that issues are resolved in a timely manner
- Ensuring adequate staffing to operate units safely
- Ensuring adequate funding to maintain equipment and safety systems in good condition

Once the COO systems are developed, management must engage the front-line supervisors and foremen to help implement and maintain them. The implementation of the COO systems is the OD portion of the process. In Chapter 3, this book offers advice on ways to overcome the initial resistance to any change in the historic ways of doing business. In Chapter 7 it also suggests ways to reward workers for ongoing commitment to maintaining high levels of operational discipline.

This book is of value to anyone who will be involved in COO/OD activities because it explains what the organization hopes to achieve and why their participation and support is crucial to overall success. Individuals in the organization will recognize the need for setting up specific processes and procedures and then strictly following them.

> COO/OD applies to critical work activities of management, employees, and contractors in all departments, not just those of the operations department. It applies every time a worker performs a task throughout the life of a facility or an organization, because it is an ongoing commitment to reliable operations. For example, quality control tests must be performed accurately and reported promptly so that the process can be kept under control.

- Management and executives will understand that their behavior and personal discipline set the standards for the entire organization.
- Technical personnel will understand why it is important to design equipment so that it is easier to operate and maintain.
- Operators will understand why it is crucial that field readings be checked against panel readings.
- Maintenance workers will understand the importance of reliably performing tasks such as routine testing and housekeeping.
- The human resources group will understand their role in fitness-for-duty, progressive discipline, salary, bonus, and retention decisions.
- Support groups, such as information technology, will understand why their support of operations and maintenance is critical to their success.

> The facility manager and the facility management team must lead by example for the system to achieve success.

The goal is for everyone to understand how reliable execution of their tasks is essential for the success of the organization.

1.4 DEFINITIONS

This section includes key definitions used throughout this book. A complete listing of definitions can be found in the Glossary.[3]

> **CONDUCT OF OPERATIONS DEFINITION**
> The embodiment of an organization's values and principles in management systems that are developed, implemented, and maintained to (1) structure operational tasks in a manner consistent with the organization's risk tolerance, (2) ensure that every task is performed deliberately and correctly, and (3) minimize variations in performance.

- COO is the management systems aspect of COO/OD.
- COO sets up organizational methods and systems that will be used to influence individual behavior and improve process safety.
- COO activities result in specifying how tasks (operational, maintenance, engineering, etc.) should be performed.
- A good COO system visibly demonstrates the organization's commitment to process safety.

---

[3] Current process safety-related definitions can also be found on the CCPS Web site.

OPERATIONAL DISCIPLINE DEFINITION
The performance of all tasks correctly every time.

- OD is the execution of the COO system by individuals within the organization.
- OD refers to the day-to-day activities carried out by all personnel.
- Individuals demonstrate their commitment to process safety through OD.
- Good OD results in performing the task the right way every time.
- Individuals recognize unanticipated situations, keep (or put) the process in a safe configuration, and seek involvement of wider expertise to ensure personal and process safety.

Table 1.2 provides examples of COO and OD issues that apply to a variety of situations.

PROCESS SAFETY CULTURE DEFINITION
The common set of values, behaviors, and norms at all levels in a facility or in the wider organization that affect process safety.

- It is possible to have a good culture for occupational safety but a less successful culture for process safety, particularly if the latter aspect does not receive focused attention.
- Different groups within an organization can have different process safety cultures.
- Process safety culture can often be observed in the behaviors that personnel exhibit when they believe that no one is watching them. Process safety culture can also be described as "the way we do things around here" in relation to process safety activities.
- Process safety culture is influenced by (1) organizational factors and (2) factors that are internal to the individual. COO focuses on the first factor while OD focuses on the second. Arguably, culture can also be affected by factors outside the organization (e.g., regulations, economic conditions, social mores), but a strong COO/OD system maintains the culture within the organization despite outside influences.

TABLE 1.2. Examples of COO and OD Issues for Various Situations

Situation	Examples of COO Issues*	Examples of OD Issues
Repair a pump	• Ensure that the work permit process is functioning properly • Ensure that workers are trained in safe work procedures • Use qualified maintenance workers • Ensure that correct repair parts and tools are available in stores (e.g., through an integrated maintenance work order system) • Reinforce good housekeeping practices • Implement maintenance systems (including labeling and lighting)	• Properly isolate the pump from process piping and power sources prior to starting the work • Understand the effects of the work on other work and interfacing systems • Follow work permit procedures and ensure that contract workers also comply • Properly check completed work • Maintain proper housekeeping • Communicate the status of repair work to operations
Start up a unit	• Ensure that operating procedures adequately address startup hazards • Identify any special issues related to the causes of the prior shutdown that might require additional attention – use the change management process where appropriate • Assess any nonfunctional safety systems or process equipment and either ensure that it is repaired or confirm that alternative measures and safeguards are effective • Properly communicate any necessary changes to the startup team in writing • Empower the operator to abort startup if required to resolve safety issues	• Use repeat-backs for all communications • Follow standard procedures and note any management instructions for modifications to the procedure • Properly log the startup sequence in the shift log or in special startup documentation • Identify deviations during startup that do not match the startup procedure, and consult with supervisors as to the correct response • Terminate the startup if safety issues are not resolved or personnel are unsure of how to proceed • If a team is involved, cross check activities with other team members to ensure that the correct sequence is followed

TABLE 1.2. Examples of COO and OD Issues for Various Situations

Situation	Examples of COO Issues*	Examples of OD Issues
Change shifts	• Establish a formal communications protocol for handover between shifts, including time to review logs • Clearly define the expected nature of communications among supervisors, board operators, and field operators • Establish a safety interlock defeat log and ensure that the logs are reviewed at the start of each shift • Establish a printed log form suitable for shift handover, rather than relying on operator notes	• Arrive promptly for shift change to allow time for adequate shift handover, and do not depart until the handover is complete • Properly log important information for the handover – process conditions, work underway, any safety equipment or interlocks out of service, etc. • Jointly review log forms transferred between the two shifts
Upgrade a level instrument	• Formalize the change management process and the forms to be completed by personnel • Assess the training needs of personnel that will arise as a result of the change	• Involve engineers, operators, and maintenance personnel when addressing all issues of concern associated with the change • Complete management of change procedures and all pre-startup assessments prior to using the equipment
Conduct the weekly plant staff meeting	• Establish a general agenda for the meeting so that personnel can be prepared for each meeting • Establish a schedule for the meeting • Track action items that result from the meeting • Assign adequate resources and completion dates for action items	• Attend meetings regularly • Review action items that are past due • Stick to the agenda and schedule • Prepare appropriate meeting notes

*Note: To avoid repetition, all COO activities include system aspects such as Planning, Implementing, Monitoring, and Management Review.

According to Merriam-Webster's dictionary (Ref. 1.9), the term "discipline" can have the following meanings:

1. punishment
2. a field of study
3. training that corrects, molds, or perfects the mental faculties or moral character
4. (a) control gained by enforcing order, (b) orderly or prescribed conduct or pattern of behavior, (c) self control
5. a rule or system of rules governing conduct or activity

The word "discipline" as used in OD does NOT refer to punishment.

Process safety risk-related OD focuses on definitions 4(b) and 5: orderly conduct and behavior and system governing conduct. Certainly one of the goals of an OD system is to establish order using a prescribed pattern of behavior. It does this through a system of rules that govern the performance of tasks in the facility and hold personnel accountable for their behavior. Trusting people to do their jobs, holding them accountable for their failings, and rewarding them for their behaviors are key aspects of a COO/OD system.

However, no set of rules or procedures can anticipate every possible situation and circumstance. Therefore, OD does not require or encourage blind compliance with any set of rules or procedures. OD encourages "thoughtful compliance" (Ref. 1.8).

Personnel are expected to follow the rules and procedures. However, **personnel are also expected to think** about what will happen if the established rules and procedures are applied to the current situation. If they believe the risks of implementing the rules and procedures are unacceptable, they are expected to stop and seek advice from other knowledgeable people. It may be possible to change the situation so that it is safe to proceed. Otherwise, they should work through the organization's process to change the rules or procedure prior to executing the modified procedures. Rules and procedures should not be changed in an uncontrolled manner. However, if an emergency requires an immediate response, then knowledgeable personnel should be trusted and empowered to enact modified procedures as a last resort to protect safety, based on their training and experience.

An example of the "thoughtful compliance" approach in emergency situations is the U.S. Nuclear Regulatory Commission's (NRC's) rules for nuclear power plant operators. Licensed plant operators are required to follow all of the conditions of their operating license and technical specifications (operating limits). However, the NRC also has a rule [10 CFR 50.54(x) (Ref. 1.10)] that states:

> *A licensee may take reasonable action that departs from a license condition or a technical specification in an emergency when this action is immediately needed to protect the public health and safety and no action consistent with license conditions and technical specifications that can provide adequate or equivalent protection is immediately apparent.*

In other words, commercial nuclear power plant operators are required to follow all the rules, except when following the rules in an emergency situation will result in unacceptable risk (i.e., endangering the public health and safety). Thus, a training and competency system that explains the "why" behind the rules is essential to support good OD.

As the effectiveness of the COO/OD system increases, the need for traditional discipline practices should decrease.

There should be appropriate traditional discipline systems to hold personnel accountable for their actions. These systems are a backup to the COO/OD process and are outside the scope of this book. However, the human resources discipline system should follow COO/OD principles in treating everyone fairly and administering the same discipline for a rule or safety principle violation. In an organization with an effective COO/OD system, managers seldom refer personnel to the human resources discipline system unless they are intentionally or recklessly endangering others. When individuals are formally disciplined, people throughout the organization generally support the decision because they refuse to tolerate willful dangerous acts on the part of their coworkers.

In an organization with an effective COO/OD system, personnel work together to encourage appropriate behaviors and discourage inappropriate behaviors through rewards and penalties integrated into work routines. As a result, the use of traditional human resources methods for disciplining people in an effort to correct their behavior is seldom required. Personnel monitor each other's performance and provide positive and negative feedback to other personnel in an effort to continuously improve the group's performance. However, when an individual's behavior makes it necessary, then the organization must take the appropriate disciplinary actions to retain its credibility.

1.5 HOW TO USE THIS BOOK

This book is organized so that readers can focus their attention on specific topics, depending on their role.

If you are just getting started with COO/OD, you should find all of the chapters helpful. If your organization's management is already supportive of COO/OD and you are just looking for specific actions to implement, focus on Chapters 5, 6, and 7.

Chapter 2 discusses the advantages and expected outcomes of implementing a COO/OD system. Chapter 3 describes the actions that leadership needs to perform to establish an effective system. Chapter 4 outlines key aspects of human factors that affect the implementation of a COO/OD system. Chapters 5 and 6 provide details on the implementation of the COO and OD systems. Finally, Chapter 7 describes the Plan-Do-Check-Adjust process associated with implementing a COO/OD system. Table 1.3 lists the range of people for whom this book was written and suggests those chapters that the authors feel would be most beneficial. A "P" indicates a chapter of primary interest to the group, and an "S" indicates a chapter of secondary interest.

1.6 HOW DO I KNOW IF I NEED TO IMPROVE MY COO/OD SYSTEM?

This section provides checklists to help organizations gauge where they are with respect to COO/OD systems. The checklists are Indicators of Effective COO/OD Systems (Table 1.4), Examples of COO System Characteristics (Table 1.5), and Examples of OD System Characteristics (Table 1.6).

If a COO/OD system is working well, most of the positive indicators in Table 1.4 should be evident, and the system would qualify for Stage 5 maturity as described in Section 7.5.3. Table 1.5 provides examples of COO system strengths and weaknesses. Table 1.6 addresses the same content for OD systems. If you see the symptoms of weakness described in the second column of these tables, COO/OD system improvements could move the performance toward what is described in the third column of the tables.

If you determine that your organization has some of the symptoms listed in Tables 1.5 and 1.6, the remainder of this book will help you identify a path for improvement.

1.7 BASIC COO/OD CONCEPTS

Figure 1.1 shows a process safety pyramid or triangle, where the minor, serious, and catastrophic injuries normally found progressing up to the top of a personal safety triangle have been replaced with appropriate process safety issues, consistent with the process safety focus of this book. Eliminating the issues at the base of the triangle should result in a reduction in process safety incidents. COO/OD activities are typically focused on the bottom portion of the triangle with the goal of reducing the number of issues that occur at higher levels of the triangle.

TABLE 1.3. Key Chapters for Each Job Position

	1 – What Is COO/OD and How Can I Tell if I Need It?	2 – Benefits of COO/OD	3 – Leadership's Role and Commitment	4 – The Importance of Human Factors	5 – Key Attributes of Conduct of Operations	6 – Key Attributes of Operational Discipline	7 – Implementing and Maintaining Effective COO/OD Systems
Executive	P	P	P				S
Plant/Facility Manager	P	P	P		S		P
Site Operations Leaders/Area Managers	P	P	P	P	P	P	P
Environmental, Health, and Safety/ Process Safety Managers/Specialists	P	P	P	P	P	P	P
Site Foreman/Front-line Supervisors	P	S	P	P	P	P	S
Engineers/Project Managers	P		S	P	P		
Operators	P			P	P	P	
Maintenance	P			P	P	P	
Laboratory Technician	P			S	P	P	
Construction Workers	S					P	
Purchasing	P			P	P		
Warehouse	P				P	P	
Human Resources	P	S	S	S	P	P	

P – Chapter of primary interest, S – Chapter of secondary interest

TABLE 1.4. Indicators of Effective COO/OD Systems

Equipment is properly designed and constructed	☐ Operational, maintenance, safety, and environmental considerations are all addressed in the initial design of equipment. ☐ Proactive risk analysis results and industry standards are used as inputs to the design process. ☐ End users of the equipment (generally operations and maintenance personnel) are involved in the design process. ☐ The design process occurs in a controlled manner. ☐ The construction occurs in a controlled manner.
Equipment is properly operated	☐ The proper method for operating equipment has been developed through proactive analysis of the risks and documented in written procedures. Operators are involved in the development of the procedures. ☐ Personnel have been trained in normal and abnormal operations, as well as the basis for the procedures and operating limits. ☐ Equipment is configured and operated in accordance with procedures. ☐ Equipment is returned to service using a controlled process. ☐ Changes to operational requirements are appropriately assessed.
Equipment is properly maintained	☐ Equipment is maintained in accordance with predetermined maintenance strategies developed through a structured assessment process. ☐ Personnel are trained to troubleshoot, repair, and maintain equipment. ☐ Changes to operational conditions are assessed to determine their impact on maintenance requirements. ☐ Equipment status is controlled through safe work practices. ☐ Equipment failures are analyzed to prevent similar failures.
Management systems are properly executed	☐ Management systems are developed based on the results of proactive analyses and industry best practices. ☐ Management systems are clearly documented. ☐ Management systems are executed as written. ☐ Organizational changes are assessed to determine impacts on existing management systems.
Errors and deviations are consistently addressed	☐ The personnel in the system are always seeking to improve their performance. As a result, there is extensive use of self-checking, peer-checking, audits, incident investigations, management reviews, and metrics to identify and eliminate deviations. ☐ Personnel are actively seeking discrepancies and resolving issues when identified. ☐ Personnel take ownership of issues and seek to solve the problem themselves. They involve outside resources to assist them in solving the problems, but retain ownership of the issue. ☐ Personnel embrace feedback from personnel outside their group as opportunities to improve their systems and processes.

TABLE 1.5. Examples of COO System Characteristics

Topic	Indicators of Weakness	Indicators of Strength
Management Systems		
	☐ Management reviews of operations are not conducted or are conducted in an informal manner.	☐ Managers have a structured management review process for process safety elements and key operation/maintenance activities and generate actions to address the issues that are identified.
COO Foundations		
	☐ A new policy or procedure is issued and the first question asked about it is: "OK, I understand what the paper says, but what do you really want us to do?"	☐ Policies describe the behaviors that are expected of personnel.
	☐ Horseplay[4] is tolerated.	☐ The consequences of horseplay and other willful or egregious acts are documented.
	☐ The schedule of maintenance tasks was known to be unrealistic due to the reduced size of the maintenance staff. However, no risk prioritization was performed to identify the most critical tasks. As a result, technicians were told to "do what they could."	☐ Required resources are provided to complete the scheduled work or the scheduled work is prioritized to be consistent with the available resources.

[4] Rowdy or boisterous play; gay or light-hearted recreational activity for diversion or amusement. Also known as skylarking.

TABLE 1.5. Examples of COO System Characteristics

Topic	Indicators of Weakness	Indicators of Strength
	☐ Managers frequently met to identify problems and corrective actions. However, because the corrective actions were not effectively tracked, most of the actions were not completed.	☐ Corrective actions are tracked to completion by the organization. ☐ Corrective actions that are past due are periodically reviewed, and actions are taken to complete the corrective actions.
People		
	☐ Logs are incomplete or inaccurate. During incident investigations, they are of no help in determining what was going on in the facility.	☐ Expectations for keeping logs in a timely manner are enforced. ☐ Logs provide sufficient detail to describe facility operations.
	☐ They let us use our judgment on using safe work practices. If we feel the risk is low, they told us we don't have to use them – "I was only in the confined space for a couple of seconds, so there is no need for all that paperwork and equipment. I just held my breath."	☐ Personnel are trained on safe work practices. ☐ Workers understand the consequences of deviating from safe work practices. ☐ The process for obtaining work permits runs smoothly to reduce the effort required to obtain the right permits. ☐ Personnel are expected to use the proper system even if the system slows the work pace. ☐ Personnel are not punished for apparent delays caused by using the proper safe work practice system.
	☐ Broken equipment, old containers, and trash are found around the facility. ☐ Spills are not controlled or investigated.	☐ Equipment is kept clean so that process upsets and leaks are easily detectable. ☐ Tools and spares are located in specific locations. ☐ Trash is promptly disposed. ☐ Obsolete equipment is promptly removed.

TABLE 1.5. Examples of COO System Characteristics

Topic	Indicators of Weakness	Indicators of Strength
	☐ Personnel are assigned to tasks based on whomever is available, regardless of their qualifications to perform the specific task.	☐ Supervisors are aware of who is qualified to perform each task. ☐ Personnel are assigned to tasks based on their qualifications to perform the specific task, not just their general job title.
	☐ Personnel crowd into the control room, making it difficult for operators to communicate with each other.	☐ Limits on control room access during startups and shutdowns are enforced. ☐ The layout of the control room allows interaction between operators and maintenance staff without compromising operations.

TABLE 1.6. Examples of OD System Characteristics

Topic	Indicators of Weakness	Indicators of Strength
Organizational Aspects		
	☐ The operations manager tells operators to adhere to procedures, but when they get in the way of performing a rapid startup, he tells them to "do what it takes to get it done."	☐ Leadership follows the same rules they preach for front-line personnel. ☐ Leadership gathers and considers input from front-line personnel when making changes to the organization/facility. ☐ Leaders do not tolerate deviations.
	☐ Workers have no real input into the design and development of procedures, training, equipment, policies, and tools. "*Those guys* keep sending us this stuff. Why do we have to follow *their* procedures?"	☐ Front-line workers provide suggestions on improvements to the management systems, equipment, procedures, and tools used in the facility. ☐ Management acts on the suggestions. ☐ Management rewards workers who suggest and help implement improvements.

TABLE 1.6. Examples of OD System Characteristics

Topic	Indicators of Weakness	Indicators of Strength
	☐ Personnel generally follow procedures. But when there are conflicts between the procedure and production, they take shortcuts to get the job done.	☐ A system of structured methods for changing procedures, from informal to formal, is in place and widely used. Each change to the procedure is assessed, using a graded approach, before it is approved. ☐ Correct procedure use is enforced. ☐ There is visible evidence of thoughtful compliance. ☐ If the procedure cannot be followed, the activity is stopped until the procedure is properly changed or an exception is approved. ☐ Management communicates its rationale for exceptions or changes to established procedures so that workers understand the situations.
	☐ Broken equipment, old containers, and trash are found around the facility. There is no drive by workers to keep the workplace clean.	☐ Workers drive the housekeeping process. They correct other workers who deviate from the housekeeping standards.

TABLE 1.6. Examples of OD System Characteristics

Topic	Indicators of Weakness	Indicators of Strength
Individual Aspects		
	☐ Workers do not seek out additional knowledge, skills, and abilities. ☐ Not knowing how to do the job right shouldn't hold you back from trying to do the work.	☐ Workers self-regulate their assignment to tasks. They do not perform tasks if they are not qualified.
	☐ "What's the point? You try hard, you don't try – it all works out the same anyway." ☐ "It's not my problem – someone else should fix that." ☐ Three temperature indicators all show different temperatures, but no effort is made to understand and resolve the differences.	☐ Personnel take ownership of problems and drive solutions. ☐ Personnel aggressively seek solutions to operational and maintenance issues.
	☐ Personnel do not perform peer-checking because it is viewed as a way to get other people in trouble. ☐ Personnel do not spend time assessing the hazards associated with tasks. ☐ Personnel do not ask questions about the status of equipment and activities being performed in their areas. ☐ Personnel do not recognize increases in pressure and temperature that indicate a runaway reaction. ☐ Personnel do not recognize dust accumulation in unoccupied spaces, which could be a precursor to a dust explosion.	☐ Personnel actively seek out additional information about the status of equipment and activities. ☐ Personnel seek out process deviations and assess their implications. ☐ During periods of low work activity, workers actively seek to expand their knowledge of the facility through such activities as "what-if" challenges.

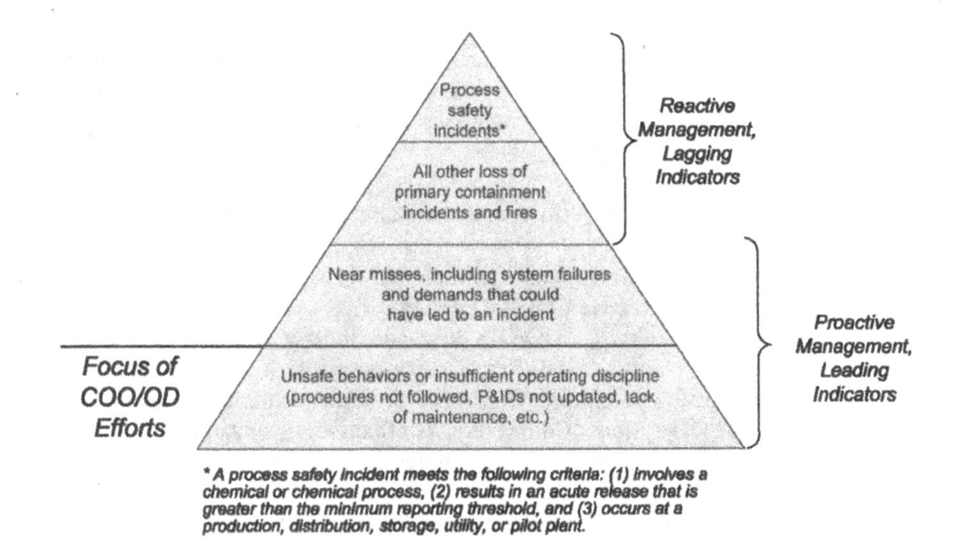

FIGURE 1.1. Typical Process Safety Pyramid

Advantages of focusing efforts near the bottom of the process safety pyramid include the following:

- Problems can be identified quickly.
  - The activities are performed frequently enough that feedback from the observations should identify potential performance gaps in a short period.
  - The undesirable and unsafe acts are leading indicators of process safety performance that can be identified and addressed before significant incidents occur.
- The activities can be easily observed.
  - The performance of most front-line workers produces an observable and measurable result (what the individual does in the field).
  - Corrective actions are handled with local resources and completed quickly. This visible response demonstrates management's commitment to COO/OD.
- Changing behaviors can change thinking.
  - For most personnel, if their behavior can be modified, their attitudes are often improved. When behaviors and attitude differ, most people will attempt to change one to eliminate the discrepancy. If the consistent performance can be maintained long enough, attitudes will usually change to embrace the new standards of performance (Ref. 1.11).

- Changes will leverage across multiple work areas/outcomes.
  - Low level behaviors, such as making rounds or completing paperwork, tend to be common to many work areas. So improvements in one area can be copied widely to improve the outcomes in other work areas.

Disadvantages of focusing efforts in this area include the following:

- The overall organizational culture may make it more difficult to implement an effective COO/OD system.
  - As noted above, COO/OD does not directly address organizational cultural issues. When working with a potentially unsafe organizational culture, it is more difficult to implement an effective COO/OD system. Conversely, an effective organizational culture will facilitate the development, implementation, and maintenance of a COO/OD system.
- PSM systems related to COO/OD may not be effective.
  - Development, implementation, and maintenance of a COO/OD system are facilitated by effective implementation of other PSM systems. If these systems have significant weaknesses, it will make implementation of the COO/OD system more difficult.
- Lagging indicators at the top of the process safety pyramid may be slow to respond.
  - Because COO/OD is focused on the bottom of the pyramid, improvements there may take months or years to demonstrably affect the top-of-the-pyramid statistics. It takes significant effort to recognize and consistently address the lower level behaviors, and consistent attention is key to reducing the base of the pyramid.
  - The legacy of poor COO/OD may be a future accident, even if the new COO/OD system is perfectly implemented.

1.8 IMPLEMENTATION OF THE COO/OD SYSTEM

Figure 1.2 outlines the basic process used to implement a COO/OD system and the corresponding chapters where each element is discussed in this book. The process can be entered from two conditions. The entry point at the top of the diagram is appropriate for a new COO/OD system (Chapter 3). The second entry point, at the bottom of the diagram, is better suited to efforts to improve an existing COO/OD system (Chapter 7). The first step for a new system is to establish (or revise) the goals and management leadership to make the system successful (Chapter 3). Next, the COO/OD systems are developed/revised (Chapters 5 and 6) and implemented (Chapter 7). As the COO/OD systems are implemented, their performance is measured (Chapter 7). Based on the performance data, revisions are made to the COO/OD system (Chapter 7). This cycle then continues as the system is monitored and improved over time. Human factors issues may arise in all

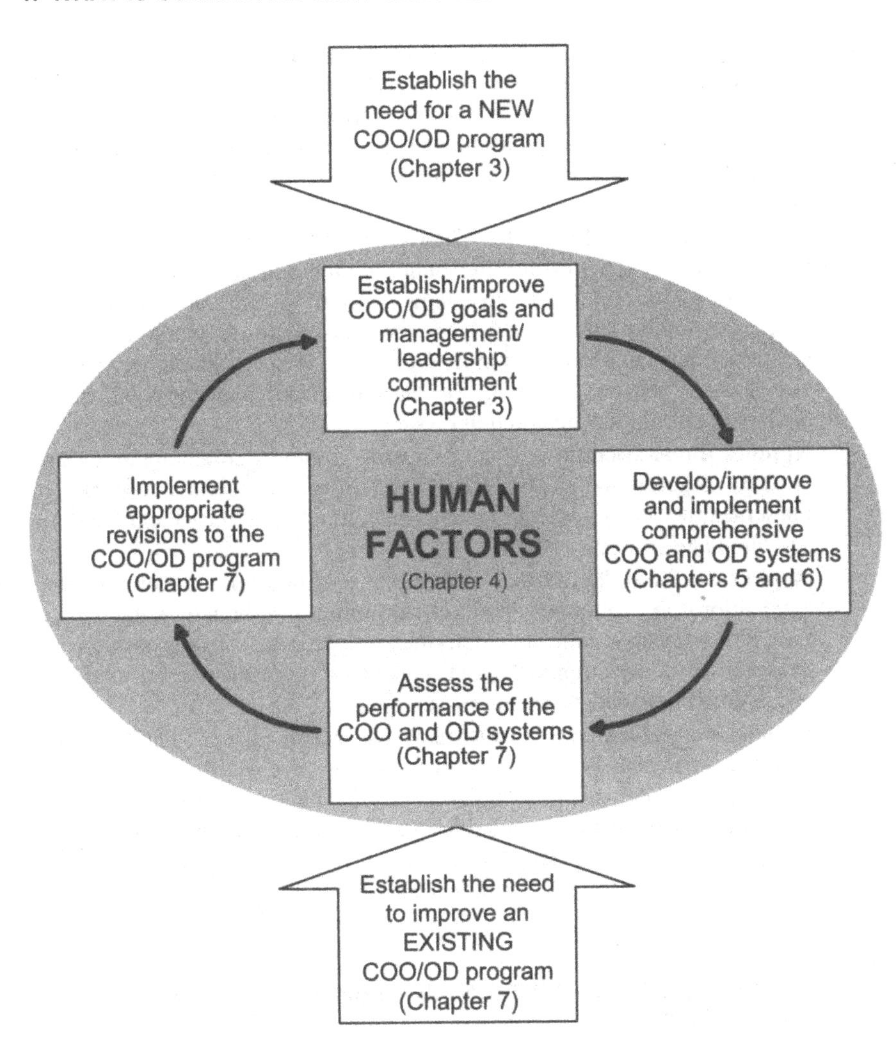

FIGURE 1.2. COO/OD Improvement and Implementation Cycle

elements of the process (as shown in the gray circle encompassing all of the elements), so they are collectively discussed in Chapter 4.

1.9 SCOPE OF THE BOOK

As discussed in Section 1.2, this book is intended to explain the key attributes of COO/OD and to provide specific guidance on how an organization can implement effective systems. Its guidance:

- **Applies throughout all levels of the organization.** The OD work activities are typically focused on the performance of front-line personnel. However, to be successful, personnel at all levels and in all functions of the organization must support and execute the process. The success of the system will be determined by the leadership provided by the facility management.
- **Applies to the total life cycle.** The COO/OD system should be applied to the entire process life cycle. Personnel involved in any aspect of the life cycle of a system (design, construction, operation, maintenance, decommissioning, demolition, and site remediation) should apply the COO/OD concepts. For example, COO/OD concepts should be used by engineering personnel during the design phase and by construction personnel during the construction phase.
- **Applies internationally.** The COO/OD system should apply to any facility, regardless of its location. However, certain aspects of implementing the system will need to be tailored to address facility culture and language issues.
- **Applies to fixed facilities.** This book was primarily developed with applications to stationary facilities in mind. Although many of the COO/OD concepts and work activities are relevant to transportation or maritime situations, application in these environments was not explicitly considered in the development of this book.
- **Focuses on process safety, not personal safety.** The COO/OD system described in this book is focused on process safety improvement. However, application of these concepts and implementation of the work activities described herein should have the added benefits of improving occupational safety, product safety, reliability, and quality, as well as reducing risks to consumers and the public.

1.10 RELATIONSHIP TO OTHER MANAGEMENT SYSTEM FRAMEWORKS

COO is closely related to several other PSM elements identified in CCPS's *Guidelines for Risk Based Process Safety* (Ref. 1.8). Foremost among them are culture, procedures (of all types), training, competency, and management review. One of the most fundamental COO requirements is consistent execution of procedures. For that to happen, (1) there must be written procedures to execute and (2) workers must be trained on the proper execution of the procedures.

Another key interfacing element is management of change (MOC). When properly implemented, OD stops a worker from improvising ways to complete a procedure when confronted with unique situations. The OD answer is to STOP as soon as safely possible, and if the situation cannot be resolved within the bounds of standard procedures and work practices, seek assistance and follow the MOC protocols.

Table 1.7 shows examples of the inputs to the COO/OD system from other elements of the RBPS structure, as well as how the outputs of the COO/OD system feed into the other RBPS elements. A table containing all of the RBPS elements can be found in the online material that accompanies this book.

> The effort required to implement a COO/OD system can be reduced by taking advantage of your existing management systems.

The COO/OD system is also related to many other commonly applied management system frameworks. Implementation of these related guidelines and regulations is anticipated to overlap with portions of the COO/OD system and reduce the effort required to implement the COO/OD system. Examples of these related guidelines and regulations include the following:

- American Chemistry Council (ACC) Responsible Care® Management System
- CCPS RBPS management system
- Control of Major Accident Hazards, U.K. Health and Safety Executive
- DSEAR – Dangerous Substances and Explosive Atmospheres Regulations, U.K. Health and Safety Executive, 2002
- ISO 9001: 2008, Quality Management Systems, International Organization for Standardization
- ISO 14001: 2004, Environmental Management Systems, International Organization for Standardization
- Occupational Health and Safety Management Systems, OHSAS 18001
- SEVESO-II, Control of Major Accident Hazards Involving Dangerous Substances, Council of the European Union, Council Directive 96/82/EC, 9 December 1996, amended November 2008
- Successful Health and Safety Management (HSG65), U.K. Health and Safety Executive, 1997
- U.K. Food Standards Agency regulations
- U.K. Offshore Installations (Safety Case) regulations 1992 SI1992/2885
- U.S. Department of Energy (DOE) Order 5480.19
- U.S. Environmental Protection Agency (EPA) risk management program (RMP) rule 40 CFR 68
- U.S. Food and Drug Administration (FDA) Federal Food, Drug, and Cosmetic Act, Current Good Manufacturing Practices
- U.S. Occupational Safety and Health Administration (OSHA) PSM regulation 29 CFR 1910.119

TABLE 1.7. COO/OD System Inputs and Outputs for Selected RBPS Elements

RBPS Element	COO – Chapter 5		OD – Chapter 6	
	Inputs	**Outputs**	**Inputs**	**Outputs**
Process Safety Culture	• Visible management support • Assessment of current process safety culture status • Process for defining safety goals • Process safety expectations • Company operational/safety philosophy	• Strengthening the process safety culture • Training programs that (1) emphasize strict adherence to procedures and practices and (2) reinforce a culture of conformance to standards • Systems for accountability for nonconforming behavior • Leadership and supervisor workshops • Establishing a questioning attitude in personnel	• Provide personnel with the authority to implement procedures and processes as designed	• Reduced process safety incidents and injuries • Questioning attitude in workers • Adherence to lines of authority • Willingness to observe and coach other workers • Intolerance of defects and deviations

TABLE 1.7. COO/OD System Inputs and Outputs for Selected RBPS Elements

RBPS Element	COO – Chapter 5		OD – Chapter 6	
	Inputs	Outputs	Inputs	Outputs
Hazard Identification and Risk Analysis	• Risk assessment methods • Risk tolerance levels • Recommendations for operating practices that will reduce the likelihood of human error or improve the effectiveness of administrative controls	• Inherently safer process characteristics • Improved hardware-related controls • Improved procedure-related controls • Improved administrative controls	• Identification of process risks and risk control measures	• Day-to-day implementation of the procedure-related and administrative controls
Operating Procedures	• Procedures that specify the appropriate steps to operate and maintain the process	• Identification of operational and maintenance issues that can be addressed through improved procedures	• Timely updates to procedures	• Implementation of the procedures • Procedural issues requiring resolution
Management of Change	• Revised process descriptions • Revised operational requirements	• Specification of trigger points for MOC assessments	• Revised procedures on which workers will be trained • Criteria for when workers must be trained versus informed	• Using the MOC process whenever it is required • Notification and review of all operational changes (e.g., bypassed alarms) • Worker training before change is commissioned

TABLE 1.7. COO/OD System Inputs and Outputs for Selected RBPS Elements

RBPS Element	COO – Chapter 5		OD – Chapter 6	
	Inputs	Outputs	Inputs	Outputs
Incident Investigation	• Recommendations for operating practices that will reduce the likelihood of human error or improve the effectiveness of administrative controls	• Receptiveness to lessons learned from investigations	• Recommendations for specific front-line personnel practices to improve performance	• Reporting of near misses • Open participation in investigations

1.11 SUMMARY

This chapter introduced the purpose of this book and defined key terms used in the rest of the book. It then described how different individuals should use the book. Example indicators of COO and OD system conditions were listed. The overall COO/OD system model was then presented.

1.12 REFERENCES

1.1 U.S. Chemical Safety and Hazard Investigation Board, *Refinery Explosion and Fire, BP Texas City, Final Report,* Report No. 2005-04-I-TX, Washington, D.C., March 2007.

1.2 Kemeny, John G., Chairman, *Report of the President's Commission on the Accident at Three Mile Island*, Washington, D.C., October 1979.

1.3 Atherton, John, and Frederic Gil, *Incidents That Define Process Safety*, Center for Chemical Process Safety of the American Institute of Chemical Engineers, John Wiley & Sons, Inc., Hoboken, New Jersey, 2008.

1.4 World Nuclear Association, "Chernobyl Accident," http://www.world-nuclear.org/info/chernobyl/inf07.html.

1.5 Cullen, The Honourable Lord, *The Public Inquiry into the Piper Alpha Disaster*, HM Stationary Office, London, England, 1990.

1.6 U.S. National Transportation Safety Board, *Grounding of the U.S. Tankership EXXON VALDEZ on Bligh Reef, Prince William Sound Near Valdez, AK March 24, 1989*, Report No. MAR-90-04, Washington, D.C., adopted on July 31, 1990.

1.7 *Final Report, Petrobras Inquiry Commission, P-36 Accident*, Rio de Janeiro, Brazil, June 22, 2001.

1.8 Center for Chemical Process Safety of the American Institute of Chemical Engineers, *Guidelines for Risk Based Process Safety*, John Wiley & Sons, Inc., Hoboken, New Jersey, 2007.

1.9 *Merriam-Webster's Online Dictionary*, http://www.merriam-webster.com/dictionary/discipline.

1.10 U.S. Government Printing Office, *Code of Federal Regulations,* Title 10, Chapter 1, Section 50.54(x), "Conditions of Licenses," Washington, D.C., revised January 1, 2003.

1.11 Cooper, Joel, *Cognitive Dissonance: 50 Years of a Classic Theory*, SAGE Publications Ltd, London, England, 2007.

2
BENEFITS OF COO/OD

2.1 INTRODUCTION

Why should an organization implement a COO/OD program? What are the potential and expected benefits of such a system? Are these benefits worth the associated investments of initial and ongoing resources? These are important questions for any organization to address prior to developing a COO/OD strategy.

2.2 OBJECTIVES OF COO/OD

As discussed in Chapter 1, COO systems are intended to encourage performance of all tasks in a consistent, appropriate manner. OD is the deliberate and structured execution of the COO and other organizational management systems by personnel throughout the organization. Much of COO/OD's history was shaped by industries (gunpowder manufacturing, nuclear weapons, aviation) that had very high potential consequences, so the benefits of risk reduction were obvious. But COO/OD is not an all-or-nothing proposition; organizations in lower risk industries can selectively adopt those elements of COO/OD that will help them achieve their goals.

Presurgery Checklist – An Example of COO/OD Benefits

In 1999, the Institute of Medicine estimated that there were up to 100,000 fatalities in U.S. hospitals each year due to avoidable medical errors; subsequent studies have estimated that the number is even higher. Medical professionals have included elements of COO in their efforts to eliminate avoidable medical errors. For example, hospitals have adopted strict procedures to avoid wrong-site surgeries (i.e., operating on the wrong leg or removing the wrong kidney).

Between October 2007 and September 2008, eight hospitals in eight countries participated in the World Health Organization's Safe Surgery Saves Lives program. The objective was to test whether implementation of a nineteen-item surgical safety checklist designed to improve team communication and consistency of care would reduce complications and deaths associated with surgery (Ref. 2.1).

Baseline data were collected on 3,733 adult patients. The surgical safety checklist was introduced, and surgical teams were trained on its use via lectures, written materials, and direct coaching. Data were then collected on 3,955 patients and compared to the baseline data.

The results clearly demonstrated the benefits that accrued from disciplined use of the checklist. Averaged across the eight hospitals, the death rate within thirty days following surgery was cut by almost half (1.5% before, 0.8% after), and the inpatient complication rate was cut by about one-third (11% before, 7% after). The individual hospitals varied widely in their economic circumstances, patient populations, and beneficial results. Not surprisingly, some of the greatest gains were made at hospitals with the worst baseline performance.

Despite his initial belief that the checklist would largely be a waste of time, one of the study's authors, Dr. Atul Gawande, voluntarily adopted the checklist in his surgical practice at Harvard so as not to be a hypocrite. Even though Harvard had a much better patient safety record than hospitals in the study group, he now admits, after using the checklist for two years, that "I have not gotten through a week of surgery where the checklist has not caught a problem" (Ref. 2.2). The checklist may institutionalize the "Hawthorne Effect," in that performance improves precisely because someone is watching (Ref. 2.3).

These results mirrored the success shown in a 2004-2005 study in 103 intensive-care units, primarily in Michigan. In that study, the use of a five-item checklist reduced the incidence of bloodstream infections from central lines up to 66%. Over 18 months, the reduction in infection rates likely saved more than 1,500 lives and $200 million in medical expenses (Ref. 2.4).

COO/OD Impact Summary

These studies reaffirmed that the benefits of COO and OD are not unique to a specific activity or industry. Instituting effective procedures and adhering to them produced the desired results – significantly improved patient outcomes in this case. The checklist reduced human errors in several critical COO/OD elements, including:

- Communication within the work group (surgical team)
- Communication between work groups (preoperative, operative, and postoperative caregivers)
- Ensuring that the right tools are available and used
- Ensuring that the right procedure is performed at the right location
- Ensuring that all required tasks are accurately completed

The concept of "benefit" must be determined within the framework of how the organization assesses "value," and in this book we are emphasizing the benefit of improving process safety. Whether pursuing profit, value addition, or efficient delivery of services, all organizations must operate within both internally and externally generated requirements (often termed "constraints" in decision support models). Examples of regulatory requirements in the U.S. include the EPA and OSHA PSM requirements and the FDA good manufacturing practices requirements. Other examples include International Organization for Standardization requirements, internal safety and quality requirements, legal and moral constraints, etc. The primary audience for this book works in or supports process industries, most of which are structured as profit-generating entities. Therefore, the following discussions focus on the concepts of value and the associated benefits of COO/OD within the framework of a for-profit business.

Figure 2.1 shows a relatively simple example of a general facility or process "value management" model. This model is based on the "Balanced Scorecard" approach of Kaplan and Norton, and it incorporates the idea that maximizing an organization's value involves more than maximizing the current quarter's profits. Kaplan and Norton (Ref. 2.5) argue that organizations can best translate their vision and strategy into results by working from four perspectives: (1) financial, (2) customer, (3) internal business processes, and (4) organizational learning and growth. In each of the four areas, the organization should establish its long-term objectives, identify appropriate measures, set near-term targets, and implement programs to achieve its goals.

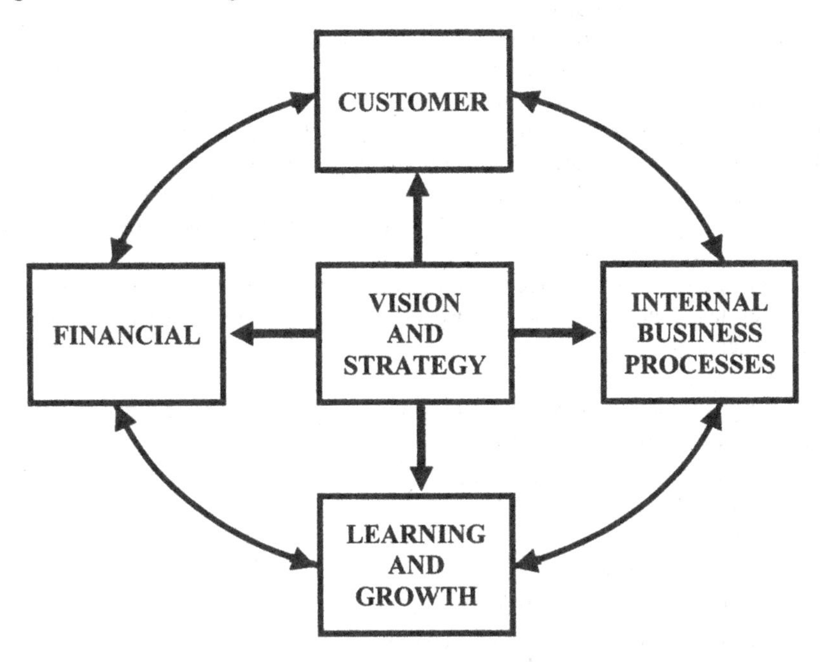

FIGURE 2.1. General Facility or Process Value Management Model (Example)

This model helps explain why modest investments in a COO/OD program can have such large and far-reaching benefits. The main motivations for embarking on a COO/OD program typically include the following:

- Improve risk management and reduce losses, waste, and downtime
- Increase product yield quantity and quality
- Improve customer service
- Improve compliance with laws, regulations, standards, and policies
- Improve the reputation of the organization

Thus, a COO/OD initiative has the opportunity to add value in all four results areas shown in Figure 2.1. For example, COO/OD reduces process safety risk, which directly improves financial results by avoiding both "ordinary" losses (e.g., the damage and downtime associated with a runaway reaction or fire) and "extraordinary" losses (e.g., the losses suffered by the community and stockholders after the accidents in Bhopal, India; Texas City, Texas; and Illiopolis, Illinois). Improving product yield and quality improves both financial results and customer satisfaction. The training required by COO/OD will also help workers learn, adapt, and improve the organization. When applied to internal business processes such as process safety management, COO/OD helps ensure not only that the organization meets regulatory requirements, but also that it does not harm its workers, damage its reputation with customers, or engender community opposition that could threaten continued operations.

Many industries and organizations have implemented systems that have strong COO/OD elements in their value chains. For example, "world-class manufacturing" is a collective term for the concepts, principles, policies, and techniques for managing and operating a manufacturing company in accordance with industry's best practices. It encompasses many of the COO/OD principles used by Japanese automobile, electronics, and steel manufacturers to drive their resurgence following World War II. It primarily focuses on continuous improvement in quality, price, delivery speed, delivery reliability, flexibility, innovation, and customer service to gain a competitive edge.

World-class manufacturing is a process-driven approach that usually involves elements such as:

- High employee involvement (see 6.2.2)
- Cross-functional teams (see 5.5.2)
- Multiskilled employees (see 5.5.11)
- Visual signaling (see 5.5.7)
- Make-to-order (see 5.7.4)
- Streamlined flow (see 5.5.6)
- Doing it right the first time (see 6.3.4)
- Total productive maintenance (see 5.7.2)
- Quick changeover (see 5.7.1)
- Zero defects (see 5.5.8 and 5.5.9)
- Just-in-time activity (see 5.5.13)
- Variability reduction (see 6.2.3)

All these elements incorporate some COO/OD concepts, but COO/OD is fundamental to doing it right the first time, achieving zero defects, completing activities just-in-time, and reducing variability.

Effective COO/OD programs enable companies to improve operations, eliminate waste, and streamline organizations. This alone results in higher productivity. But COO/OD programs also allow these companies to increase the speed of total throughput, from order capture through delivery, which eliminates heavy dependence on inventory and its associated costs. Sequential methods of performing work can be replaced with concurrent methods to compress time, and functional and hierarchical divisions of duties can be replaced by team-driven activities.

The success of COO/OD programs is also evident in the Toyota Production System, also known as Lean Manufacturing. The "basic wastes" concept was originally developed by one of Toyota's engineers, Taiichi Ohno (Ref. 2.6). He observed that products significantly differ between factories, but the typical wastes found in manufacturing environments are quite similar:

- Overproduction
- Transporting
- Unnecessary inventory
- Underutilization of employees
- Waiting
- Inappropriate processing
- Unnecessary or excess motion
- Defects

For each waste, there is a strategy to reduce or eliminate its effect on a company, thereby improving overall performance and quality. COO/OD strategies apply to all eight of these fundamental wastes. For example, COO strategies help reduce overproduction, transporting, unnecessary inventory, and underutilization of employees. OD helps reduce waiting, inappropriate processing, unnecessary or excess motion, and certainly, defects.

A good COO/OD system can be viewed as a type of "control system" for human performance. It permits appropriate variability for intelligent human control, yet achieves desired performance within established constraints (e.g., safe operating limits). Therefore, the implementation of sound COO/OD systems helps ensure that organizations maximize the value they add while conforming with requirements, which is a hallmark of organizations that succeed over the long term.

2.3 EVOLUTION OF COO/OD SYSTEMS

Throughout history, most industries and organizations have recognized that reliable human performance is essential to their success. Different management systems were developed, but the most successful ones shared many common features, which are collectively called COO/OD today. The following sections provide some historical perspective on the evolution of COO/OD systems.

2.3.1 Success in Military Applications

Since the first armed conflict, military commanders recognized that reliable human performance was the key to battlefield success. Hence they developed systems for training their troops, communicating information, and maintaining readiness for action. Each advance in military technology has placed new emphasis on human reliability, because the consequences of error have escalated. For example, an improperly aimed missile can be far more consequential than an improperly aimed cannon, which can be far more consequential than an improperly aimed arrow. As targeting failures (e.g., "friendly fire" incidents) have become more damaging, the military COO/OD systems have been continuously developed and improved. However, the development and deployment of nuclear weapons demanded a quantum improvement in organizational performance to forestall any unauthorized or accidental detonation.

Thus, the military revamped its COO/OD systems to further reduce the possibility of an uncontrolled detonation of any nuclear device, and to minimize the probability of loss of, or damage to, deployed nuclear devices. One of the best examples of an effective COO/OD system was developed by the U.S. Navy for the operation of nuclear ballistic missile submarines. These complex vessels not only have the capability to deliver large ballistic missiles carrying multiple nuclear warheads, they also are powered by seagoing nuclear power plants. When these vessels were deployed in the 1950s, Admiral Hyman Rickover was instrumental in institutionalizing an exemplary COO/OD system to maintain high standards through the life cycle of each vessel. The Navy system particularly emphasizes the development of accurate procedures and rigorous training in their use. One of the mantras of the nuclear Navy is "verbatim compliance with written procedures." The Navy's long history of no ships lost due to nuclear power plant incidents and no loss of nuclear weapons control incidents speaks for itself.

2.3.2 Success in U.S. Department of Energy Applications

As the civilian organization responsible for developing and producing nuclear weapons, the DOE and its predecessor agencies also understood their obligations for reliable performance. Incidents of grave concern to the DOE include unintentional detonation of a nuclear device, the loss of control of nuclear weapons materials, security violations involving nuclear weapons development and manufacturing processes, nuclear criticality accidents, and uncontrolled radiological releases. However, some of the companies originally hired to operate the government facilities for the DOE did not have effective COO/OD systems in place, and there were some serious incidents.

To improve organizational performance, the DOE adapted elements of military COO/OD systems and applied them to its nuclear activities. In 1990, the organization formally published its COO system for the entire organization in DOE Order 5480.19, Change 2, *Conduct of Operations Requirements for DOE Facilities* (Ref. 2.7). The purpose of this Order is "to provide requirements and guidelines for Departmental Elements ... to use in developing directives, plans, and/or procedures relating to the conduct of operations at DOE facilities. The implementation of these requirements and guidelines should result in improved quality and uniformity of operations." The DOE has implemented its COO/OD system through a series of guidance documents, including procedures, instructions, and manuals. Several of the standard DOE guides to good practices are publicly available and are listed in the additional reading section at the end of this chapter.

The DOE Facility and Administrative Support Unit has stated (Ref. 2.8) that implementation of COO/OD is helping them achieve the following objectives:

- Properly control and make readily available appropriate procedures, thereby promoting their use and helping to ensure that operational activities will be conducted in the manner intended
- Encourage appropriate management attention to writing, reviewing, and monitoring operations procedures to ensure that their content is technically correct and the wording and format are clear and concise
- Ensure that procedures that affect safety-related equipment and emergency procedures are reviewed by the appropriate facility safety review committee or by other appropriate review mechanisms
- Encourage effective review of procedures prior to issuance and at periodic intervals to ensure that the information and instructions are technically accurate; to ensure consistency in procedure format, content, and wording, which is essential to achieve a uniformly high standard of operator performance
- Ensure that procedures are developed with consideration for the human-factor aspects of their intended use
- Ensure that important factors (such as operating limits, warnings, cautions, etc.) are highlighted in procedures
- Ensure that procedures incorporate appropriate information from applicable source documents
- Ensure that referenced documents in procedures are correctly identified
- Ensure that outdated procedures are not used by mistake and that working copies are replaced
- Encourage operations activities to recognize that safety and productivity are compatible goals

The diverse and changing facilities, missions, and cultures encompassed by the DOE constantly challenge its COO/OD system. Incidents still occur, lessons are shared across the enterprise, and the COO/OD system is improved. However, the COO/OD system has successfully reduced the number and severity of incidents since its implementation.

2.3.3 Success in Aviation Industry Applications

During the first half of the twentieth century, the aviation industry expanded from primarily military applications to civil applications (the air cargo and passenger business). However, in 1947 U.S. air carriers averaged a fatal accident every sixteen days, so there was significant public distrust of the aviation industry. Thus, it was imperative that commercial airlines improve their performance to attract passengers and meet public safety expectations.

The military experience of many individuals and companies involved in commercial aviation facilitated the adoption of COO/OD principles in commercial aviation. Initially, the focus was on improving aircraft maintenance and manufacturing practices because so many accidents resulted from equipment failures. Later, the focus shifted to operational procedures, communications, pilot and crew training, and crew resource management. These efforts have proven spectacularly successful. From 2001 to 2006, there was a period of more than 1,600 days with only 1 fatal commercial airline accident in the U.S. From 1947 to 2009, the fatality rate dropped from near 50 to less than 0.05 fatalities per billion passenger-miles flown (Refs. 2.9 and 2.10). So despite the occasional tragic failure, the result of comprehensive COO/OD systems in the aviation industry is an enviable transportation safety record per passenger-mile of service.

2.3.4 Success in Utility Industry Applications

The commercialization of nuclear energy followed much the same trajectory as aviation in its conversion from military to civilian use. In his "Atoms for Peace" speech in 1953, President Dwight D. Eisenhower envisioned harnessing atomic power for the benefit of mankind. However, many in the electric utility industry viewed nuclear reactors as just a new source of heat, like coal- or gas-fired boilers, and they failed to appreciate the need for a high-reliability organization to use it successfully.

Initially, the problem was primarily reliability. The power that was going to be "too cheap to meter" was not cost competitive because the plants had so many unplanned shutdowns and outages. However, the 1979 accident at the Three Mile Island plant demonstrated that the consequences of a nuclear accident that threatened public safety could be financially devastating, both to the owner and to the industry.

Thus, the Institute of Nuclear Power Operations (INPO) was established in 1979 to promote the highest level of safety and reliability – to promote excellence – in the operation of nuclear electric generating plants. One of its earliest objectives was to implement or improve management systems for the conduct of operations at the plants. As various utilities implemented the model programs offered by INPO, the value of the programs became evident (Ref. 2.11). The average number of scrams (rapid shutdowns) of nuclear reactors dropped by 80% from 1980 to 1990 (7.4 to 1.6 scrams per reactor per year). The number of safety system actuations dropped by 60% in the five years from 1985 to 1990 (2.74 to 1.05 actuations per reactor per year). The average nuclear power plant's capacity factor (the fraction of its theoretical maximum power output actually produced) has risen from about 60% in the 1970s to more than 90% since the year 2000. Individual plants routinely set records with continuous runs of more than 700 days and capacity factors near 100%.

The successful implementation and maintenance of effective COO programs has been one of the key factors in preventing accidents and making nuclear energy one of the least expensive ways to generate electrical power. Given their success in nuclear applications, many utilities are implementing COO/OD systems to improve the reliability of other power-generating and distribution assets.

2.3.5 Success in Process Industry Applications

Chemical plants, refineries, and many other process facilities routinely use and/or create hazardous materials. These processes often involve controlled reactions of chemicals at high pressures and temperatures. Plant managers have long recognized the need for reliable human performance to minimize the risks of runaway reactions, fires, explosions, or releases of hazardous materials. However, industry practices varied widely, depending on the risk tolerance of the organization.

The E. I. du Pont de Nemours and Company (DuPont) was founded in 1802 with a core value for understanding and managing the hazards associated with its processes (Ref. 2.12). From its beginnings manufacturing gunpowder on the Brandywine River, DuPont recognized that human errors could have severe safety and economic consequences. DuPont documented its safety "rules," so they were available as a standard reference for everyone. However, explosions that occurred at the Brandywine facility in 1815 and 1818 underscored the need for disciplined adherence to the rules. Lessons from the early years of the company's history are today reflected in many elements of its current process safety program, including:

- Conduct of operations strategy
- Safe work practices
- Training
- Incident investigation
- Pre-startup safety reviews
- Emergency response
- Operational discipline

From explosives to chemicals to its many diverse businesses of today, DuPont has achieved superior safety and financial performance for more than 200 years by instituting COO/OD systems at its facilities worldwide.

However, few others in industry made such a complete commitment to COO/OD. As the size and complexity of industrial complexes grew, so too did the potential consequences of human errors and organizational failures. After a series of major accidents in Flixborough, England; Seveso, Italy; and Bhopal, India, the process industries came to the same realization as the nuclear utilities – the consequences of major accidents extend far beyond the specific facility involved. This prompted trade associations such as the ACC and the American Petroleum Institute (API), and professional organizations such as the American Institute of Chemical Engineers' (AIChE's) CCPS, to actively help their members improve process safety performance.

For example, Responsible Care® is a global initiative currently supported by more than fifty national associations, which share a common commitment to advancing the safe and secure management of chemical products and processes. Participation in Responsible Care® is mandatory for ACC member companies, all of which have made CEO-level commitments to uphold these program elements:

- Measuring and reporting performance
- Implementing the Responsible Care® Security Code
- Applying the modern Responsible Care® Management System to achieve and verify results
- Obtaining independent certification that a management system is in place and functions according to professional standards

Regulatory organizations also imposed new requirements on facilities that manufacture, store, or use hazardous materials. After the Flixborough accident in 1974, the U.K. Health and Safety Executive took the lead in developing the first goal-oriented regulations for process safety. In 1982, the European Union adopted Council Directive 82/501/EEC on the Major-Accident Hazards of Certain Industrial Activities (commonly called the Seveso Directive), and in 1992, OSHA implemented regulatory requirements for the *Process Safety Management of Highly Hazardous Chemicals* (29 CFR 1910.119).

Thus, many companies have turned to COO/OD to help them meet industry and regulatory requirements in addition to their own objectives. Like DuPont, the Dow Chemical Company (Dow) had a long history of effective COO/OD implementation. In 1964, Dow first published a Fire and Explosion Index (Ref. 2.13) that recognized OD among the loss control credit factors. Specifically, "... a fully documented operating discipline [system is] an important part of maintaining satisfactory control of a unit." Dow's current COO/OD system is embodied in three primary areas as outlined below:

1. Common Management System
    a. Leadership Responsibility
    b. Planning
    c. Implementation and Operation
    d. Checking and Corrective Action
    e. Management System Review
2. ACC Responsible Care® (Environmental, Health, and Safety [EH&S])
    a. Community Awareness and Outreach
    b. Distribution and Logistics
    c. EH&S Engineering/Design and Control
    d. Emergency Preparedness and Response
    e. Employee Health and Safety
    f. Non-Company Services
    g. Pollution Prevention
    h. Process Safety
    i. Product Stewardship
    j. Security

3. Facility Operations
    a. Operational Reliability
    b. Process Control
    c. Process Information
    d. Process Technology
    e. Operating Facilities
    f. Produce to Plan/Record Production Data
    g. Empowerment

At Celanese Corporation (Celanese), the COO/OD system is also designed to support its commitment to the highest standards of safety, personal conduct, and business integrity around the world (Ref. 2.14). Via implementation of its COO/OD system, Celanese has met its internal targets for:

- Making safety a precondition for everything they do
- Communicating openly and honestly in every situation
- Proactively safeguarding themselves, others, and the environments in which they do business
- Adhering to the highest standards of business ethics and personal conduct

Given the successes demonstrated by companies with formal COO/OD systems, the CCPS identified COO as an essential element of a comprehensive RBPS management system. The *Guidelines for Risk Based Process Safety* (Ref. 2.15) identified twenty RBPS elements and organized them into four pillars of process safety (Commit to Process Safety, Understand Hazards and Risk, Manage Risk, and Learn from Experience). The COO element is included in the Manage Risk pillar. Chapter 17 of the RBPS guidelines outlines the key principles and essential features of the COO element and lists more than fifty possible work activities related to the element (with associated implementation options), examples of ways to improve the effectiveness of the element, metrics, and management review activities related to the element.

2.4 SUMMARY

A strong COO/OD system helps maximize an organization's value to all of its stakeholders. COO/OD focuses on improving human performance – as an individual and as an organization – by establishing clear expectations and consistently performing tasks in accordance with those expectations. In other words, it helps establish a culture in which deeds match words. Many different industries and organizations have implemented COO/OD systems and have universally seen their performance improve as a result. The proven benefits of effective COO/OD systems include:

- Reduced frequency of undesired events such as injuries, material releases, fires, explosions, unplanned outages, and quality defects
- Reduced consequences associated with undesired events over the organization/facility/process life cycle
- Heightened sense of purpose, value, motivation, and well-being among organization team members and stakeholders
- Increased and sustained high levels of profitability, safety, quality, and value addition for businesses
- Increased and sustained high levels of service, safety, quality, and value addition for government agencies and not-for-profit organizations
- Increased and sustained high levels of productivity for organization team members
- Increased and sustained high levels of PSM system effectiveness
- Consistently high levels of procedural awareness and compliance

2.5 REFERENCES

2.1 Haynes, Alex B., M.D., M.P.H., et al., "A Surgical Safety Checklist to Reduce Morbidity and Mortality in a Global Population," *The New England Journal of Medicine*, Massachusetts Medical Society, Waltham, Massachusetts, Vol. 360, No. 5, January 29, 2009, pp. 490-499.

2.2 Gawande, Atul A., M.D., M.P.H., "Checklist for Surgery Success," National Public Radio interview, January 5, 2010.

2.3 Gawande, Atul A., M.D., M.P.H., et al., "Correspondence – A Surgical Safety Checklist," *The New England Journal of Medicine*, Massachusetts Medical Society, Waltham, Massachusetts, Vol. 360, No. 22, May 28, 2009, pp. 2372-2375.

2.4 Pronovost, Peter, M.D., Ph.D., et al., "An Intervention to Decrease Catheter-Related Bloodstream Infections in the ICU," *The New England Journal of Medicine*, Massachusetts Medical Society, Waltham, Massachusetts, Vol. 355, No. 26, December 28, 2006, pp. 2725-2732.

2.5 Kaplan, Robert S., and David P. Norton, *The Balanced Scorecard: Translating Strategy Into Action*, Harvard Business School Press, Boston, Massachusetts, 1996.

2.6 Ohno, Taiichi, *Toyota Production System: Beyond Large-Scale Production*, Productivity, Inc., Portland, Oregon, 1988 (English translation of *Toyota seisan hōshiki*, Diamond, Inc., Tokyo, 1978).

2.7 U.S. Department of Energy, DOE Order 5480.19, Change 2, *Conduct of Operations Requirements for DOE Facilities*, Washington, D.C., October 23, 2001.

2.8 Collins, S. K., and F. L. Meltzer, *Document Control and Conduct of Operations*, Idaho National Engineering Laboratory, EG&G Idaho, Inc., Idaho Falls, Idaho, 1993.

2.9 Air Transport Association of America, Inc., *Statement before the Aviation Subcommittee of the Senate Commerce Committee*, Washington, D.C., April 10, 2008.

2.10 International Civil Aviation Organization, "2006 International Civil Aviation Day Focuses on Aviation Safety and Security to Preserve Growing Benefits of Air Transport," Montreal, Canada, December 1, 2006.

2.11 International Research Council, Commission on Engineering and Technical Systems, *Nuclear Power: Technical and Institutional Options for the Future*, National Academy Press, Washington, D.C., 1992, p. 56.

2.12 Klein, James A., "Two Centuries of Process Safety at DuPont," *Process Safety Progress*, American Institute of Chemical Engineers, New York, New York, Vol. 28, Issue 2, June 2009, pp. 114-122.

2.13 American Institute of Chemical Engineers, *Dow's Fire & Explosion Index Hazard Classification Guide, 7th Edition*, John Wiley & Sons, Inc., Hoboken, New Jersey, 1994.

2.14 Celanese Corporation, *Sustainability at Celanese 2008*, www.celanese.com/sustainability_report_2008.pdf.

2.15 Center for Chemical Process Safety of the American Institute of Chemical Engineers, *Guidelines for Risk Based Process Safety*, John Wiley & Sons, Inc., Hoboken, New Jersey, 2007.

2.6 ADDITIONAL READING

- DOE-STD-1030-96, *Guide to Good Practices for Lockouts and Tagouts*, 1996.
- DOE-STD-1031-92, *Guide to Good Practices for Communications*, 1992.
- DOE-STD-1032-92, *Guide to Good Practices for Operations Organization and Administration*, 1992.
- DOE-STD-1033-92, *Guide to Good Practices Operations and Administration Updates through Required Reading*, 1992.
- DOE-STD-1034-93, *Guide to Good Practices for Timely Orders to Operators*, 1993.
- DOE-STD-1035-93, *Guide to Good Practices for Log keeping*, 1993.
- DOE-STD-1038-93, *Guide to Good Practices for Operations Turnover*, 1993.
- DOE-STD-1039-93, *Guide to Good Practices for Control of Equipment and System Status*, 1993.
- DOE-STD-1040-93, *Guide to Good Practices for Control of On-shift Training*, 1993.
- DOE-STD-1042-93, *Guide to Good Practices for Control Area Activities*, 1993.

- Gawande, Atul, *The Checklist Manifesto: How to Get Things Right*, Metropolitan Books, Henry Holt and Company LLC, New York, New York, 2009.
- Institute of Nuclear Power Operations, *Guidelines for the Conduct of Operations at Nuclear Power Stations*, Atlanta, Georgia, 2001.
- Institute of Medicine, Committee on Quality of Health Care in America, Linda T. Kohn, Janet M. Corrigan, and Molla S. Donaldson, eds., *To Err is Human: Building A Safer Health System*, National Academy Press, Washington, D.C., 2000.
- Rees, Joseph V., *Hostages of Each Other: The Transformation of Nuclear Safety Since Three Mile Island*, The University of Chicago Press, Chicago, Illinois, 1994.
- Zhan, Chunliu, M.D., Ph.D., and Marlene R. Miller, M.D., M.Sc., "Excess Length of Stay, Charges, and Mortality Attributable to Medical Injuries During Hospitalization," *The Journal of the American Medical Association*, American Medical Association, Chicago, Illinois, Vol. 290, No. 14, October 8, 2003, pp. 1868-1874.

3
LEADERSHIP'S ROLE AND COMMITMENT

3.1 INTRODUCTION

Management commitment is the first step toward implementing an effective COO/OD system. However, unlike other management initiatives, this one cannot simply be assigned a budget and delegated to others for execution. Yahoo! CEO Carol Bartz observed, "There is a real difference between managing and leading . . . Managing winds up being the allocation of resources against tasks. Leadership focuses on . . . help[ing] people succeed." Truly effective COO/OD systems start at the top with upper management demonstrating the behaviors they want others to emulate. This chapter describes the enduring leadership commitments required to initiate and maintain effective COO/OD systems, and the steps an organization might take to improve its COO/OD systems. This chapter should be read by managers who are contemplating whether to add or revitalize COO/OD as an element of their overall safety management system and by those who will be developing it.

As described in the *Guidelines for Risk Based Process Safety* (Ref. 3.1), COO is one of the twenty recommended elements in a comprehensive PSM program – the whole of which requires senior management attention. COO is tightly intertwined with other RBPS elements, such as operating procedures and management of change, and adherence to COO (i.e., operational discipline) both reflects and affects the organization's culture. Organizations rely on OD to ensure the effectiveness of engineered and administrative barriers to prevent incidents. Thus, OD drives system reliability, and COO drives OD; therefore, upper management must drive COO to achieve the overall safety performance desired.

3.2 ACHIEVING GREATNESS WITH COO/OD

In his book entitled *Good to Great* (Ref. 3.2), Jim Collins describes his team's research findings on the common characteristics of good companies that transformed themselves into great companies. He describes the transformation as a process of buildup, followed by breakthrough, with three distinct stages:

1. Disciplined People
2. Disciplined Thought
3. Disciplined Action

Piper Alpha – An Example of the Importance of Leadership Commitment

Piper Alpha was a North Sea oil production platform operated by Occidental Petroleum (Caledonia) Ltd. (Occidental). The platform began oil production in 1976 and was later modified to also produce gas. An explosion and resulting fire destroyed it on July 6, 1988, killing 167 people. Total insured loss was about £1.7 billion (US$3.4 billion). To date it is the world's worst offshore oil disaster in terms of lives lost.

Description of the Accident

The disaster began during routine maintenance activities. A pressure safety valve (PSV) on a backup propane condensate pump needed to be tested while the pump itself was isolated for an overhaul. The testing could not be completed by 6:00 p.m., so the workers were given permission to finish the work the next day. The open flange was covered with a plate, but the delay was not communicated to the operations staff.

The Piper Alpha Platform Before the Accident

Later in the evening (about 9:55 p.m.) during the next work shift, the primary condensate pump failed. None of those present were aware that the PSV maintenance work was incomplete, and the two work permits did not cross-reference each other. Believing that the planned work had not yet started, workers reversed the electrical isolation of the backup pump and started it. However, workers could not see the missing PSV, and the cover plate could not contain the high-pressure gas.

The escaping gas ignited at about 10:00 p.m. The explosion ripped open firewalls, allowing the fire to spread, and soon large quantities of stored oil were burning out of control. The automatic water deluge system had been temporarily disabled to protect divers from being accidentally drawn into the pump intake, but no one had been stationed at the firewater pump controls. When the fire broke out, there was no accessible means to manually start the system.

Flame impingement weakened gas risers from other platforms that continued production despite Mayday calls from Piper Alpha. These steel pipes, 24 to 36 inches in diameter, contained flammable gas at 2,000 psig. About twenty minutes after the initial explosion, the risers burst and dramatically increased the size of the fire.

The accommodation block was not smoke-proofed, and conditions got so bad that some people decided that the only way to survive would be to escape the platform immediately. However, smoke and flames blocked all routes to the lifeboats and helipad. In desperation, they jumped into the cold seawater hoping to be rescued by boat. Sixty-two workers saved themselves this way; most of the other 167 died from carbon monoxide and fumes in the accommodation block.

COO/OD Impact Summary

The Cullen report on Piper Alpha was highly critical of Occidental's management, stating that management commitment was "superficial" and led to poor practices and ineffective audits. This caused both ineffective COO and OD.

The Piper Alpha Platform After the Accident

The poorly designed permit-to-work process reflected the weak COO/OD system. Failure to cross-reference active permits relying on the same electrical and mechanical lockouts made the system vulnerable to communication failures.

Poor communication was a key root-cause contributor to this disaster – a classic example of ineffective OD. This was particularly evident in (1) the failure of maintenance to notify operations that the PSV testing was incomplete, (2) the failure of maintenance to notify operations that the firewater pumps were disabled, (3) the inadequate escape and evacuation instructions to Piper Alpha workers once the event began, and (4) the inadequate emergency communications between upper management in Aberdeen, platform operations, and onsite emergency responders.

Failure to stop production on interconnected platforms accelerated and worsened the accident. Lacking upper management's overt commitment to safety in COO and unable (at 10:00 p.m.) to get upper management's permission to shut down, managers on adjacent platforms erred on the side of production and continued to pump oil and gas into lines connected to the Piper Alpha platform, rather than shutting down to help depressurize the lines.

Regular safety audits were not properly conducted. When a major problem was found, it was often ignored. By such inaction, management clearly signaled its lack of interest in COO and thus few, if any, problems were ever brought up. (Ref. 3.3)

Essentially, the organization's leadership must commit itself to the transformation, develop appropriate and effective management systems, and ensure that those systems are rigorously implemented. Collins used financial results as his criteria for "greatness," but most of the principles also apply to those who want to transform their organization to produce great process safety results as well.

3.2.1 Disciplined People

Collins describes Level 5 leaders as those who blend extreme personal humility with intense professional will. Others call them "servant" leaders. "Level 5 Leadership" is the first and most basic requirement for the conversion of a good company to a great one. Managers contemplating a COO/OD system often begin with the dream that their organization could be great, *if only* their people would get things right. Sadly, those managers are doomed to be disappointed because they believe that COO/OD applies to their underlings – **not** to them personally.

The ambition of Level 5 leaders is directed toward the success of the organization, the workers, and the community, not their personal glory. Thus, Level 5 leaders realize that for the organization to achieve greatness, they must personally embrace and diligently follow the tenets of COO/OD that they espouse to others. They, in turn, see COO/OD as the pathway to success in the organization, modify their own behavior, and influence others. The cascading effect quickly spreads COO/OD throughout the organization, from the boardroom to the shop floor. Level 5 leaders overtly demonstrate their commitment to process safety, one of the four "pillars" of RBPS. As the Chinese general Sun Tzu said, "Leaders lead by example and not by force."

Unfortunately, most people do not want to change. Things are "good enough" just as they are, and there is no need for a long and arduous journey. Thus, an immediate challenge for visionary leaders is to decide who they want to join them on the journey. There are undoubtedly many good people in the organization who simply need a little persuasion to push them out of their comfort zone. Explaining the organization's objectives, explaining how COO/OD can help achieve those objectives, demonstrating management commitment to COO/OD, and rewarding those who demonstrate success through COO/OD will enroll many of the doubters. However, some may be unwilling or unable to embrace the concepts of COO/OD and apply them daily. These people, despite their talent or past contributions, must be persuaded to participate or be replaced; otherwise, they will sabotage the entire endeavor.

> When the U.S. Navy converted to nuclear-powered submarines, there was a clique of veteran sailors with a "diesel boats forever" attitude. However, the ingenuity with which they kept the old diesels running was diametrically opposed to the operational discipline required in the nuclear Navy. Those who refused to adopt and embrace OD were retired, transferred, or left ashore.

Buckingham and Coffman (Ref. 3.4) also stress the importance of selecting the right people and assigning them to the right roles. Both they and Collins come to the same, somewhat pessimistic, conclusion that people do not change much, so the wrong people cannot be remolded to implement a COO/OD strategy that is different from their personal value system. Thus, a key to long-term success is to indoctrinate new employees in the COO/OD system so that it becomes part of their expectations, and that those expectations are reinforced over time so that COO/OD is an integral and successful part of their work experience.

Once people with the right knowledge, skills, and abilities have been selected and positioned in the organization, they are the engine that will drive implementation of the COO/OD system. They sincerely believe that COO/OD offers a route to greatness, and they are willing to commit themselves to it. However, COO/OD is not a "one size fits all" system. Certainly, there are basic principles that must pervade the organization, but specific systems for various work areas and work groups will inevitably vary in detail. Nevertheless, having committed people decide how to best implement COO/OD in their work areas is the surest path to success.

The ultimate success of a COO/OD system requires management's enduring commitment. COO/OD will not be an overnight success, and there will be many setbacks along the way. Thus, it is usually best that management not announce the introduction of COO/OD with great fanfare and grandiose goals.. A better approach is to begin simply. Establish some modest, achievable goals with a system to measure progress. Apply COO/OD principles and let the results speak for themselves.

Level 5 leaders will focus on improving their own efforts rather than on whom they can blame the failure. Implementing a successful COO/OD system requires overcoming the massive inertia of an organization. In accordance with Newton's First Law of Motion, a body at rest tends to remain at rest. Thus, management's initial efforts to get COO/OD underway may produce few outward signs of improvement. Yet, the initial applied force produces some movement, and some successes are observed.

Management must then build on those initial successes by showing employees how COO/OD benefitted both individual workers and the organization as a whole. People generally want the financial and emotional rewards associated with being part of a winning team, so a record of success entices others to join the program and commit real effort to its further success. Lessons learned from OD failures (including tragic accidents, such as the loss of a colleague) can also be used to galvanize the workforce's efforts to implement COO/OD. In either case, Newton's Second Law takes over and the implementation of COO/OD begins to accelerate in direct proportion to the force applied. Collins refers to this as the "flywheel effect" and acknowledges that the overall success of the program reflects the cumulative effect of everyone's effort. Management's commitment is a necessary step, but it alone is not sufficient to achieve the organization's objectives.

3.2.2 Disciplined Thought

Disciplined thought begins with an attainable goal and unwavering dedication to its achievement. Will COO/OD significantly contribute to the fundamental economic drivers in the organization? If not, it is doubtful that the organization can maintain the long-term commitment required to attain and sustain effective implementation. However, money alone is unlikely to capture the hearts and minds of front-line workers who see only a distant relationship, if any, between the efforts they must expend versus the financial reward.

Thus, management needs to be committed to the long-term implementation of COO/OD as a means to achieve something about which the organization is deeply passionate. Management must define the organization's core values to start the process of disciplined thought. Those thoughts will become actions, and repeated actions will become habits in the performance of everyday activities. Elite athletes use a similar process to visualize and train for record-breaking performances. For example, many organizations have embraced a goal of zero incidents that result in harm to workers, the public, or the environment. If this is a reflection of their core values, then COO/OD offers a practical means to achieve measurable progress toward that goal as well as their economic goals. However, if espousing such a goal is mere sloganeering, efforts to implement COO/OD will likely fail.

Disciplined thought is not wishful thinking. Management must be willing to confront the reality of the organization's current situation. What was it that drove the organization to consider implementing COO/OD? A severe downturn in its economic fortunes? A quality problem that tainted its reputation in the marketplace? A significant incident that killed workers or neighbors? Management's post-traumatic reaction to such significant emotional events is to grab for something, *anything*, that might solve the current crisis and, ideally, prevent its recurrence. Unfortunately, COO/OD is not a "quick fix" program. The crisis may provide the initial impetus to get COO/OD underway, but it will require sustained management effort, as previously discussed, to implement and maintain it.

Disciplined Thought

1. Define the goal
2. Commit to its achievement
3. Appraise the current situation
4. Confront reality
5. Repeat until the goal is attained

Hence, an objective appraisal of the organization's current situation is required. Maybe there has been a long history of labor/management strife, and the workers suspect that COO/OD will just be another management tactic to blame them for any problems. Maybe the company is strapped for cash and is cutting back on everything. Maybe the company prides itself on a long history of doing whatever it takes to get the job done under any circumstances. It is only by discovering and accepting the facts of the current situation that management can confront reality and work toward a different future.

If management truly embraces the process of disciplined thought, then reality becomes the starting point for improvement plans to reach the organization's goals. If labor/management strife has been the history, then the COO/OD system will particularly need to address how each employee will be held to the same standards. If financial constraints are the reality, then the initial COO/OD plans will have to focus on areas that cost little to implement, such as improving communication and housekeeping. If getting the job done is the reality, then the COO/OD plan must include a change management system that balances the risks and rewards without being overly burdensome.

"Our situation is a fact, but it does not define our future. We are willing to do whatever is necessary for as long as necessary to achieve our goal of greatness. We will commit ourselves to a series of large and small steps that will help us achieve our goal, over the long term, regardless of any setbacks along the way." Such statements indicate that management is willing to hear the unvarnished truth and develop plans that move the organization toward its goals. Such candor will produce more good decisions over the long run and will facilitate the successful implementation of COO/OD.

Confronting harsh truth might appear to be counterproductive, but this is not so. If the right people have been put in key positions, candor will be refreshing, not discouraging. These people will be stimulated by the challenge, particularly when the boss acknowledges the difficulty of achieving the goal. Anything less will be perceived as naïve, deceitful, or patronizing.

3.2.3 Disciplined Action

As a culture of operational discipline pervades the organization, it sets the stage for breaking through to great performance. This is where COO/OD really pays off. Many companies evolve into large, unwieldy, expensive bureaucracies in reaction to incompetence and undisciplined behavior. In contrast, OD is a commitment to disciplined behavior in executing procedures. Managers and workers implicitly trust each other to fulfill their commitments and expect to be held accountable if they fail to do so. Thus, a truly effective COO/OD system does not hinder innovation and efficiency – it enhances them. (See the NUMMI example in Chapter 7.) Disciplined people are self-starters who do not need micro-management, much less the burden of an enormous, and stifling, bureaucracy. Not only is such bureaucracy expensive, it also tends to drive off the best workers. Disciplined people prefer to be treated as adults, not children.

Disciplined thinking focuses on solving problems. Disciplined action frees upper management to focus on ways to capitalize on business opportunities rather than on ways to impose tighter controls. The objective of disciplined thought and action is to encourage creativity and responsibility within an established framework. For example, a hospital establishes the framework for delivery of medical services. Doctors are expected to work within that scientific, ethical, and legal framework, yet they are free to prescribe whatever they believe will be the most effective treatment regimen for each individual patient. Everyone (doctors, nurses, orderlies, cooks, administrators, etc.) has responsibilities, not just jobs. When their collective decisions and actions produce superior health outcomes, the hospital develops a reputation as a great medical institution.

However, discipline alone is not guaranteed to produce great results. Disciplined execution of a flawed plan is futile. Rigorous monitoring of average corrosion rates will not preserve mechanical integrity if pitting or cracking is the dominant failure mode. The goal is to have disciplined people engage in rigorous thinking and then take action within a formal system. A self-sustaining culture of discipline empowers employees to use their skills and knowledge to do their jobs well every day. This helps the organization sustain itself and overcome adversity.

3.3 LEADERSHIP'S ROLE IN INSTITUTING COO/OD

COO/OD offers many safety, environmental, and economic benefits to an organization. However, the primary focus of this book is on improving process safety performance, so we will focus on management's role in that regard.

The purpose of any COO/OD system is to reliably accomplish the mission of the organization within the framework of its core values. Thus, upper management's most fundamental responsibilities are to:

- Identify values consistent with the organization's mission
- Translate values into business principles
- Establish policies consistent with those principles
- Communicate policies throughout the organization
- Consistently uphold business standards and practices based upon them.

Implementing COO/OD for process safety purposes requires that upper management state how the organization values the safety of people and the quality of the environment so that those values can be embedded in the business standards and practices that guide individual actions.

This immediately poses a dilemma because no organization exists simply to be safe. Any human activity involves some element of risk. Setting a goal of zero human injuries is admirable, but if the organization values human life above all else, the possibility of a traffic accident would prohibit employees from traveling to work or delivering goods and services. Maintaining a safe workplace, protecting the community, and/or minimizing impacts on the environment allow the organization to achieve its mission on a sustained basis. Confronting the reality that organizations do accept some risk to provide benefits to their employees, communities, and investors requires that upper management set realistic risk tolerance criteria that the organization can achieve. A parallel commitment to continuous improvement (i.e., risk reduction) then sets the organization on a path to great safety and environmental performance over the long term.

Thus, implementing a COO/OD system (as described in Chapter 7) is a pragmatic way for management to ensure that the organization's values are embedded in its daily activities. Even if good management systems already exist, OD is essential for good performance (i.e., following the procedures that comprise the good system). Improving OD positively affects all aspects of the business (quality, reliability, profitability, reputation, occupational safety, environmental impact, etc.) as well as process safety.

3.3.1 Clearly Define Expectations

Paraphrasing the Cheshire Cat in *Alice's Adventures in Wonderland*, "If you don't know where you're going, any road will take you there." Like Alice, management wants to go SOMEWHERE, and the Cheshire Cat reassuringly says, "Oh, you're sure to do that ... if you only walk long enough."

Trying to implement a COO/OD system without clearly defined expectations is much like Alice's trip through Wonderland. To avoid merely arriving somewhere, management must define clear expectations for the system. Otherwise, each stakeholder has a different idea of what defines success and in which direction the finish line lies. Like a group of children asked to draw a dog, each imagines an entirely different animal, from a Chihuahua to a St. Bernard. Management must create a vision of success – what it looks like, feels like, sounds like, tastes like, and smells like. Every single employee must share a similar vision about the end result to which the organization is committed so that all can work toward that common goal. It is not an easy process, but it is essential.

First, management must decide what it wants the organization to achieve (THE VISION). The vision is an articulation of the desired future state of the organization – what it aspires to become. Does it have to be realistic? The vision may be very difficult to achieve, but it needs to be viewed as possible if the organization fulfills its mission statement. Management next has to determine where the organization is now (REALITY). This requires hard work, research, and self-examination. The best approach employs some systematic appraisal of the problem (e.g., accident investigations, culture surveys, town hall meetings, or possibly audits). These findings can then be fed into a structured OD assessment, such as the management system maturity model discussed in Section 7.5.3.

Finally, management must build a path backwards from the goal to today's reality. This establishes the expectations – SMART goals that will serve as identifiable mileposts en route to accomplishing the mission (for example, reducing the number of overdue maintenance inspections by 10% by the end of this year). SMART goals help people convert the grand plan into a series of manageable steps that they can accomplish.

SMART Goals

- ✓ Specific
- ✓ Measurable
- ✓ Attainable
- ✓ Relevant
- ✓ Time-specific

The expectations must reasonably address known obstacles. For example, it would be absurd to expect workers to immediately begin following procedures if no procedures exist, or if existing procedures are wrong or outdated. So a short-term goal would be to get the procedures for one work group written or updated and validated within six months. Workers in that business unit can then be trained and begin more disciplined adherence to those procedures while other units are writing or updating their procedures. Inevitably, problems will arise that were not anticipated. A key person may leave, economic conditions may change, or a major accident may occur. If the goals are truly important to the organization, management must remain committed and persevere like a ship's captain encountering a storm at sea. The expectation is that the cargo will be delivered. He may have to navigate around the storm or batten down the hatches and ride it out, but he will get through the storm.

The expectations must be accompanied with a clear set of priorities because there will be times when workers must resolve conflicts among complex, multifaceted goals. Is it more important to deliver the cargo on schedule or to protect the ship? The answer seems obvious, but the wrecks of the RMS *Titanic* and the *Torrey Canyon* (Ref. 3.5) resulted from the schedule being given priority over the safety of the vessel. A similar commitment to schedule over safety played a key role in the 1986 explosion of the Space Shuttle Challenger. As management states its expectations regarding the implementation of COO/OD, it must be clear on which expectations take precedence. This hierarchy must be restated on a regular basis and reinforced by actions to be effective.

3.3.2 Clearly Define Acceptable Limits

The goal of OD is for there to be no deviations from the established COO system, as described in Chapter 5. It is, therefore, essential that there be a practical definition of "deviation," which is unacceptable, versus "variation," which is normal, expected, and tolerable. Acceptable limits should be defined to ensure that the worker's results are reasonably fit for purpose, not that they conform to some arbitrary standard of perfection. For example, specifying that log entries shall be made with a black pen and reprimanding those who deviate by using blue ink is a mistake. Management will be perceived as being more interested in enforcing nitpicky rules than in getting work done safely and efficiently. On the other hand, requiring workers to complete checklists as tasks are performed, but tolerating individuals who leave the checklists blank, is also a mistake because it leaves compliance to the whims of individuals. Between those extremes are situations where workers inadvertently learn that acceptable product results when a batch is held at temperature for only twenty minutes rather than the thirty minutes specified in the procedure. In that case, management should enforce the operational discipline of submitting an MOC request and getting approval before modifying the hold time. Failure to do so encourages the "normalization of deviance" and undermines the entire COO/OD system. The key is to define deviations much like safe operating limits – exceeding them puts the organization's goals at risk.

When unusual situations arise, the COO system should have a prescribed mechanism for dealing with them. For the front-line workers, the most commonly implemented approach is "stop work" authority. Workers are told that if they do not feel they can do a job safely, they should stop the job in a safe state and get their supervisor and/or safety department involved in resolving the issue. For COO application, this authority is usually expanded to include a requirement for workers to exercise "thoughtful compliance" and stop work whenever they encounter a situation that cannot be resolved with standard procedures and practices. For example, an operator might run out of an essential batch ingredient. At that point, the worker should involve others and follow standard management of change procedures to address the unique situation.

Another approach is to use risk-based criteria in defining acceptable limits. The concepts of risk are well established as a means to guide consistent decision-making across diverse activities and business units within an organization (Refs. 3.1 and 3.6). Management can define the organization's risk tolerance in a matrix form, as illustrated in Figure 3.1. The higher the likelihood or severity of a mishap, the higher the risk it poses to the organization. When the front-line workers cannot achieve tolerable risk with the available equipment and current procedures, they must seek the approval of higher levels of management before undertaking, or continuing, higher risk activities. The higher the risk, the higher the management level required for approval. This approach helps ensure that deviations from the norm are approved by someone with the authority to accept that risk for the organization, or to commit the resources needed to lower the risk if he or she does not deem it tolerable. For example, the operations manager could accept the risk of continuing operation with a bypassed interlock, or order that the unit be shut down until the interlock is repaired.

	Low Severity	Medium Severity	High Severity
High Frequency	Marginal Risk	High Risk	High Risk
Medium Frequency	Tolerable Risk	Marginal Risk	High Risk
Low Frequency	Tolerable Risk	Tolerable Risk	Marginal Risk

FIGURE 3.1. Example Risk Matrix

3.3.3 Consistently Enforce Expectations

Implementing a COO/OD system may require fundamental changes in an organization's culture. Beyond simple inertia, management must overcome Newton's Third Law – for every action there is an equal and opposite reaction. Management should expect resistance to change, particularly if the front-line workers perceive that the change is, or could be, a threat to them. Unfortunately, the threat does not have to be real or large to provoke resistance. Most workers have experienced the "flavor of the month" initiative, and they believe that this, too, will pass. If their current situation is familiar and relatively comfortable, then there is no real reason for them to change. Moreover, even if they believe that COO offers real benefits to the organization, they do not trust that management will follow through.

Management's only effective tool in countering this resistance is the consistent enforcement of its expectations at all levels of the organization. The key is to enforce expectations regarding *behavior*, not outcome, and this is particularly important in organizations where the word "discipline" has traditionally meant punishment of those blamed (usually the front-line worker) for an incident. For COO/OD to succeed, management must treat the worker who achieves record production rates by bypassing safety interlocks as seriously as the worker who causes an incident by bypassing the same interlocks. In both cases, the behavior violates COO/OD principles, even though the two outcomes are opposite. Similarly, managers who encourage, tolerate, or knowingly ignore unacceptable behavior are as guilty of violating COO/OD principles as the perpetrator of the behavior, and they deserve equal treatment. In healthcare and aviation, the term "just culture" (Ref. 3.7) is used to describe OD systems that are fundamentally fair in satisfying demands for individual accountability while contributing to organizational learning and improvement.

> A supervisor and her group were initially rewarded for exceeding production goals. Later, management discovered that the production goals were exceeded by shortcutting some required procedural checks. Although there were no process incidents during the manufacturing process, the company withdrew the reward because the group had not adhered to company requirements. The reward was subsequently given to a group that met production goals AND adhered to company requirements.

3.3.4 Monitor Performance Data

Identifying and using relevant metrics are key to monitoring COO management system effectiveness and providing input to continuous improvement. A combination of leading and lagging indicators is usually the best way to provide a complete picture of system effectiveness. Outcome-oriented lagging indicators, such as incident rates, are only useful if "incidents" (i.e., losses due to failures of the COO system) occur fairly frequently. As a result, in mature COO systems, leading indicators, such as the rate of improperly performed line-breaking activities or improperly bypassed alarms/interlocks, are more useful.

Lagging indicators, such as the number of injuries per labor hour, per operating hour, or per pound produced, can also distort relative performance. A highly automated facility producing millions of pounds may be a company's best performer per pound produced, but its worst performer per labor hour. However, measuring performance as percentage attainment of such simple ratios may produce anomalous results. A group missing a single inspection may look bad at 0% attainment, while a group missing 100 of 1,000 inspections may look relatively good at 90% attainment.

The frequency for refreshing the individual metrics may range from daily to weekly to monthly or longer, depending upon the dynamic nature of the metrics, the anticipated costs of data collection, and the local needs. The objective is to select a set of metrics that are sensitive enough to help facility management monitor the performance and efficiency of the management system on a near-real-time basis without overwhelming management with a data dump. Properly chosen metrics enable managers to:

- Identify evolving management system weaknesses
- Make adjustments to work activities before the activities degrade into a failed state (performance or efficiency)
- Reinforce and maintain good practices and performance

A variety of possible COO/OD metrics, such as the number of improper permits issued or the number of overdue inspections, are discussed in Chapter 7. In many cases, existing metrics can be used for COO/OD purposes. The CCPS, the API, and the U.K.'s Health and Safety Executive have published guidelines specifically for process safety metrics (Refs. 3.8, 3.9, 3.10, and 3.11). In addition, the CCPS is compiling a database of three process safety metric reports (Process Safety Incidents Count, Process Safety Incident Rate, and Process Safety Severity Rate) to help industry monitor progress toward process safety improvement goals. Even though they are lagging metrics, companies implementing COO/OD can use those data as a starting point to benchmark themselves against others in industry.

Typically, a small set of metrics is proposed, data are gathered, and the set is pilot tested to determine whether tracking the metric data helps identify management system performance. The data values and recent trends can then be compared to management's expectations, and celebrations or corrective actions can be initiated, as appropriate. Management should review the metrics collected periodically, and those metrics that do not support the Plan-Do-Check-Adjust cycle (described in Section 7.1) should be eliminated.

3.3.5 Verify Implementation Status and Progress

In the beginning, frequent monitoring will be required. Management cannot simply define the expectations and blindly delegate the implementation. As previously discussed, early successes will become the foundation for ongoing gains in the COO/OD system. Management can use these successes to build momentum for further implementation of COO/OD. Conversely, early failures, if undiagnosed and untreated, will be particularly harmful to the overall health of the system.

In the *Guidelines for Risk Based Process Safety* (Ref. 3.1), the CCPS introduced the concept of management review as an essential management system to complement traditional auditing. Management review is the routine evaluation of whether management systems are performing as intended and are efficiently producing the desired results. Regular management reviews fill the gap between day-to-day supervisory activities and formal periodic audits. Management review is intended to be a more timely, and less formal, review of management systems so that incipient problems can be spotted and resolved before a system failure occurs. COO, like any other management system, will also benefit from regular management reviews.

The depth and frequency of monitoring should be governed by factors such as the current life-cycle stage of the facility, the maturity (degree of implementation) of the COO/OD system, the level of management performing the review, and past experience (e.g., incident history, previous reviews, and audit results). Monthly monitoring is appropriate for a new or substantially changed system, but monitoring will likely transition to semiannual or annual management reviews as the system matures. Since every level of management – from the process supervisor to the plant manager to the corporate manager – is involved in COO/OD, all should participate in periodic management reviews.

The management review checks the implementation status of the COO/OD system against established requirements. (See Chapters 5 and 6 for a discussion of specific requirements.) The management reviewer(s) invites individuals responsible for managing and executing system elements to a meeting where they present system documentation and implementation records, offer their direct observations of conditions and activities, and answer questions about system activities. The team attempts to answer questions such as:

- What is the quality of our system?
- Are these the results we want?
- Are we working on the right things?
- Can we do this more effectively?

Anticipated challenges (organizational changes, staff changes, new projects, new standards, etc.) to the COO/OD attributes described in Chapters 5 and 6 are also discussed so that management can proactively address those issues. Thus, the management review facilitates medium-term planning and fills the gap between long-term strategy and short-term tactics.

Recommendations for addressing any existing or anticipated performance gaps or inefficiencies are proposed, and responsibilities and schedules for addressing the recommendations are assigned. Typically, the same system that tracks corrective actions from audit findings also tracks management review recommendations to their resolution. The meeting minutes and documentation of the resolution of the recommendation are maintained as required to meet programmatic needs.

3.3.6 Sustain Performance

Figure 3.2 shows the role of management/leadership commitment in the overall COO/OD improvement and implementation cycle. This commitment must occur first if a new system is being put in place (entering the cycle from the top of the diagram); otherwise, it is the commitment to make a significant improvement in an existing COO/OD system (entering the cycle from the bottom of the diagram). In either case, management/leadership commitment is required before the organization embarks on developing a new system or significantly revising existing systems as described in Chapters 5 and 6. Although management/leadership commitment is shown as a discreet step, management/leadership commitment is a factor in successful implementation of all of the steps in the cycle.

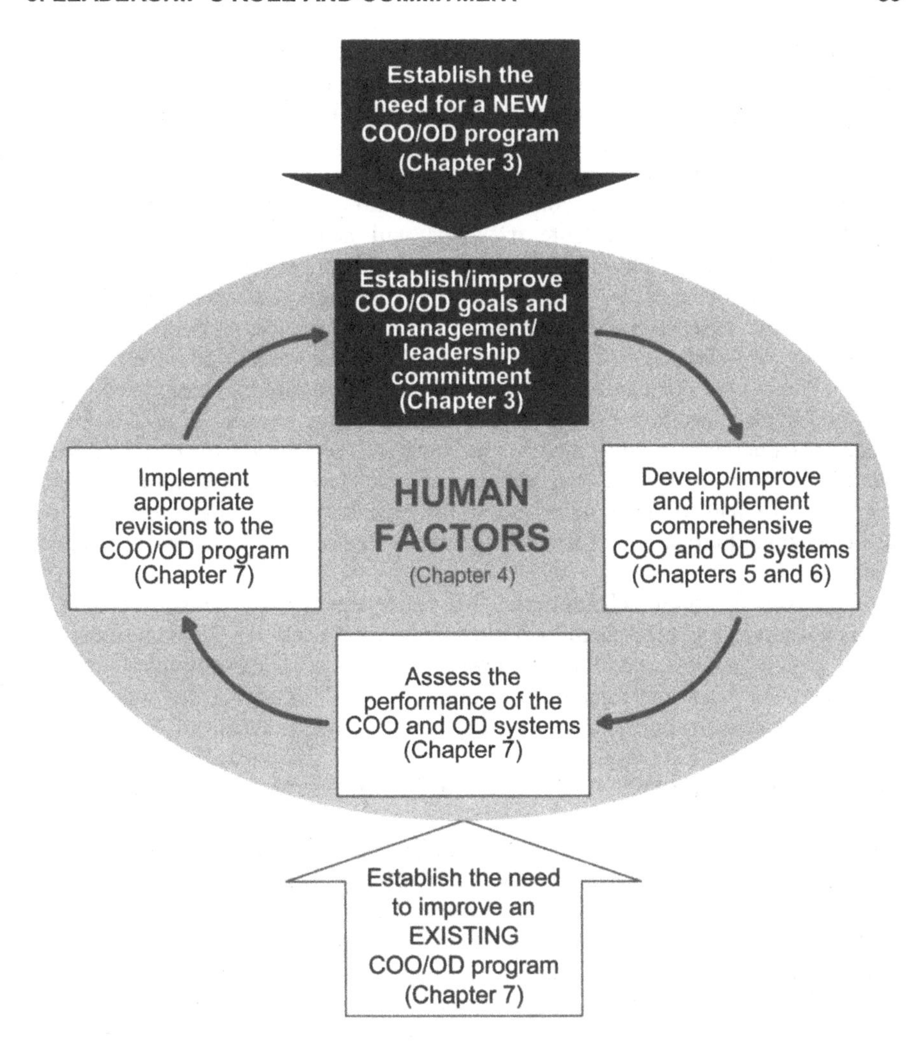

FIGURE 3.2. COO/OD Improvement and Implementation Cycle

3.3.7 Consider the Impact of a Catastrophic Event

In Section 3.2.2, we discussed the need to confront reality as an element of disciplined thought. One uniquely difficult reality could be the occurrence of a catastrophic event immediately before, or during, the early implementation of a COO/OD system.

In the immediate aftermath of the incident, there will be a significant degree of organizational paralysis. Some will be in shock at the loss of their friends and colleagues, and some will be angry or bitterly disappointed that such a thing could happen in "our" organization. Others will feel guilty, even if they were not directly at fault, and question whether they could have done something differently or better to avoid the incident. Some will be frozen by uncertainty about whether their actions will adversely affect the company during legal or regulatory proceedings. For some, their life will focus on understanding what happened and ensuring that it will never happen again.

However, the organization cannot long survive in such a climate. Thus, it is prudent to have a plan for conducting operations during the crisis and the operational discipline to execute it. This will help overcome organizational paralysis and direct the emotional energy into useful improvements. Repairs must be made, contracts must be fulfilled, and revenue must be generated. In such circumstances, operational management duties must be assigned, at least temporarily, to individuals who are not physically or emotionally involved with the event. Those who are can be productively assigned to the incident investigation team and to planning how to avoid recurrence of similar events. If failures in the conduct of operations contributed to the incident, the emotional energy of these individuals can be harnessed to accelerate and expand implementation of COO/OD. The need for change is irrefutable, and so the organization is more receptive to the idea of change, particularly if it will help ensure that such an event never happens again.

However, this approach presumes that management fully embraces COO/OD and is exemplifying its principles during the recovery effort. If management is simply trying to do *something*, but does not embrace and exemplify COO/OD principles itself, then the initiative will quickly fail. Memory of the incident will fade, and the organization will revert to pre-incident business as usual.

3.3.8 Implement COO/OD Across a Global Workforce

Companies everywhere deliver products and services to their customers using a multicultural workforce, whether or not they are global companies. When implementing COO/OD, it is wrong to believe that what works in one location will work everywhere. While the organization should have one guiding set of values and principles, the implementation of COO/OD must be tailored to the reality of specific business units. For example, some sites may have workers in collective bargaining units with contract requirements that are different, or irrelevant, at other sites. A world-class COO/OD system is NOT simply an exemplary model superimposed on a global workforce without much regard to local differences.

Effective implementation of COO/OD requires strong leadership with cultural sensitivity and a willingness to adapt the policy to be congruent with the lawful local business practices, labor expectations, political system, social pattern, religion, and approach towards work. While the concept of building a "culture-blind," global workforce is enticing, people are not the same everywhere. It would be foolhardy to expect that a generic policy of any sort could address the complex and idiosyncratic issues of a multinational, multicultural, and/or multiregional workforce. Workers may agree with the objectives of COO/OD, but they will have different perspectives on what makes a good procedure, how best to resolve conflict, and what is important in getting the job done.

Despite these differences, the fundamental steps are still the same. Management must select people who understand the local culture and who can work productively with people who may be different from themselves. These leaders then provide the COO/OD "vision" for the workforce and define expectations. The organization's COO/OD policy statement sets the baseline for global workforce practices, but the vision should reinforce the relevance of the local culture. These leaders can also confront the reality of the needs of people from the local culture and define acceptable deviations from the global standards. Achievable goals must be established, and progress toward those goals must be monitored. Effective managers will know how to motivate the local people to achieve those goals and how to use the initial successes to entice others to embrace the principles of COO/OD. Thus, the detailed system implementation should be tailored for each region and operated, more or less, on a stand-alone basis. However, relevant performance-monitoring data should be transferred to a central database for analyzing the performance of the whole organization.

3.4 SUMMARY

To build a successful COO/OD system, upper management must develop a vision of greatness for the organization and communicate that vision to everyone in the organization. Management must make an enduring pledge of time and resources to achieve the vision and shift its focus to enlisting support and assistance from lower levels of the organization. Workers will immediately want to know:

- What is COO/OD?
- What does the organization hope to gain?
- How will it affect me?

Personnel who share upper management's passion must be put in positions of influence and authority. Expectations for worker performance across the organization must be defined and enforced. Organizational performance must be monitored, and as milestones are achieved, management should lead celebrations of that success while challenging the organization to meet its next goal on the path to greatness.

Chapter 4 addresses human factors concepts that influence individual and organizational behavior and are pertinent to COO/OD. The following two chapters describe key attributes of COO/OD systems. Chapter 7 then describes the Plan-Do-Check-Adjust cycle for implementing the COO/OD system and continuously improving it.

3.5 REFERENCES

3.1 Center for Chemical Process Safety of the American Institute of Chemical Engineers, *Guidelines for Risk Based Process Safety*, John Wiley & Sons, Inc., Hoboken, New Jersey, 2007.

3.2 Collins, Jim, *Good to Great: Why Some Companies Make the Leap . . . and Others Don't*, HarperCollins Publishers Inc., New York, New York, 2001.

3.3 Cullen, W. Douglas, *The Public Enquiry into the Piper Alpha Disaster, Vols. 1 and 2,* Stationery Office Books, London, England, 1990.

3.4 Buckingham, Marcus, and Curt Coffman, *First, Break All the Rules: What the World's Greatest Managers Do Differently*, Simon & Schuster, New York, New York, 1999.

3.5 Petrow, Richard, *The Black Tide: In the Wake of Torrey Canyon*, Hodder & Stoughton, London, England, 1968.

3.6 Center for Chemical Process Safety of the American Institute of Chemical Engineers, *Guidelines for Developing Quantitative Safety Risk Criteria*, John Wiley & Sons, Inc., Hoboken, New Jersey, 2009.

3.7 Dekker, Sidney, *Just Culture: Balancing Safety and Accountability*, Ashgate Publishing Limited, Aldershot, England, 2007.

3.8 Center for Chemical Process Safety of the American Institute of Chemical Engineers, *Guidelines for Process Safety Metrics*, John Wiley & Sons, Inc., Hoboken, New Jersey, 2009.

3.9 ANSI/API Recommended Practice 754, *Process Safety Performance Indicators for the Refining and Petrochemical Industries*, American Petroleum Institute, Washington, D.C., April 2010.

3.10 U.K. Health and Safety Executive, *A Guide to Measuring Health & Safety Performance,* London, England, December 2001.

3.11 U.K. Health and Safety Executive, *Developing Process Safety Indicators: A Step-by-Step Guide for Chemical and Major Hazard Industries*, HSE Books, London, England, 2006.

3.6 ADDITIONAL READING

- Center for Chemical Process Safety of the American Institute of Chemical Engineers, *Building Process Safety Culture: Tools to Enhance Process Safety Performance*, New York, New York, 2005.
- Walter, Robert J., *Discovering Operational Discipline*, HRD Press, Inc., Amherst, Massachusetts, 2002.

4
THE IMPORTANCE OF HUMAN FACTORS

4.1 INTRODUCTION

This chapter describes broad human factors issues, which are the organizational, environmental, and job factors that influence the behavior of personnel in the organization. The chapter defines human error and describes the categories of human error that can be used to identify solutions to human performance issues. It also examines the relationship between the COO/OD system and common human performance programs (behavior-based programs, antecedent-behavior-consequence programs, and human performance technology). These factors should be considered in the design, development, and implementation of the COO/OD system.

The relationship of Chapter 4 to the remainder of this book is shown in Figure 4.1. Human factors issues underlie all the topics covered by the other chapters, from management leadership (Chapter 3) to COO/OD system development (Chapters 5 and 6) to implementation (Chapter 7).

Chernobyl – An Example of the Importance of Human Factors

On the morning of April 26, 1986, a massive explosion lifted the roof off of Reactor Building No. 4 at the Chernobyl Power Station, near Kiev. The facility was attempting to conduct an experiment to improve the facility's response under emergency conditions. The test was scheduled to start at 1:00 a.m. on April 25. However, the test was postponed for 22 hours because of electrical power demand. By the time the plant received permission to conduct the test at about 11:00 p.m. on April 25, most of the technical staff had left the facility for a long holiday weekend, and the plant was far from the initial conditions required for the test.

Although graphite-moderated reactors can be unstable under certain conditions (causing a runaway exothermic reaction), the Chernobyl plant, and ones like it, had been successfully operated for decades. However, safe operation necessitated following certain operational policies.

Fearing that failure to perform the test because of some minor delay or procedural issues would embarrass the plant staff, the operating crew proceeded despite the absence of the technical staff or other limits in the test procedure. During the test, numerous operational policies were knowingly violated, including:

- Shutting down the emergency core cooling system
- Increasing reactor coolant flow to the core above the authorized limit

- Operating the plant at less than 1,000 MW, making the plant difficult to control at the time of the test
- Disabling the rapid reactor shutdown system
- Making on-the-spot changes in the procedures to address different initial test conditions, without performing a formal assessment of the impact of the changes

When plant personnel decided to proceed with the test despite the unstable plant conditions, the nuclear reaction began to accelerate rapidly and the operators could not react quickly enough to regain control of the reactor. A mixture of hydrogen and carbon monoxide was created from high temperature reactions of the graphite and water in the reactor. When the hydrogen ignited, the resulting explosion blew off the concrete roof of the reactor building and sent nuclear material into the atmosphere, causing increased atmospheric radiation readings as far away as Western Europe.

So why did the operating staff decide to knowingly violate numerous requirements?

- The conditions required for performing the test were difficult to create, so this was viewed as a rare opportunity to make the facility stand out.
- There would be rewards for the plant staff for completing this important test.

The strong desire on the part of the plant staff to do whatever it took to complete the test, including violating numerous administrative controls and exhibiting a lack of operational discipline, led to the worst nuclear accident in history. (Refs. 4.1 and 4.2)

4.2 HUMAN BEHAVIOR ISSUES

The following summarizes a few key issues related to human behavior that must be addressed by the COO/OD system:

- **People are fallible, and even the best make errors.** Everyone in your organization will err, even highly motivated, well-trained people using good procedures. Certainly, one COO/OD goal is to prevent errors, but another is to design processes so that errors do not result in catastrophic consequences. To prevent and mitigate human errors, organizations must plan for them by implementing prevention, detection, and correction safeguards for human errors.

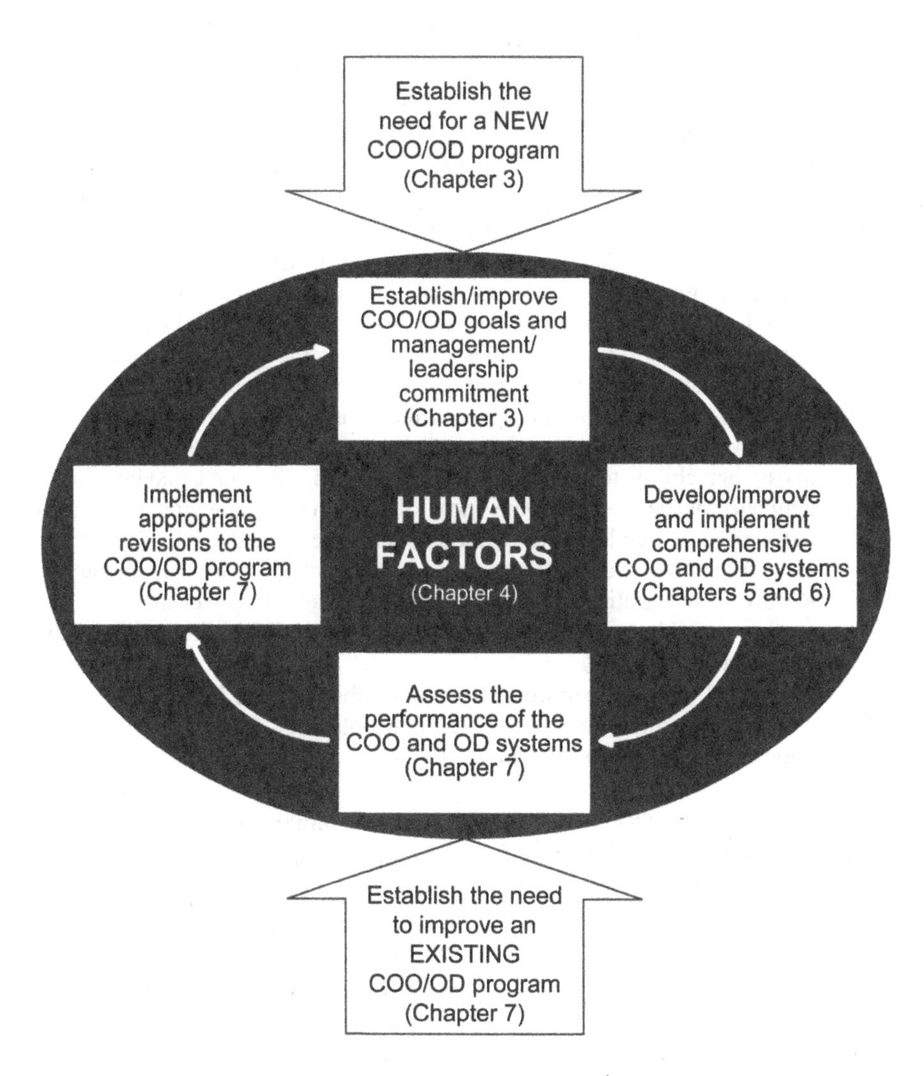

FIGURE 4.1. COO/OD Improvement and Implementation Cycle

- **Error-likely situations are predictable, manageable, and preventable.** Human errors are normally not random. Most human errors can be predicted in advance, and systems can be put in place to prevent, detect, and correct the errors so that they do not produce catastrophic consequences. Therefore, organizations must spend time identifying error-likely situations and implementing systems to manage these situations.
- **Organizational processes and values influence individuals.** Management systems are used to encourage desirable behaviors and discourage undesirable behaviors. The design and implementation of the

management systems related to equipment design, operation, maintenance, procedures, supervision, incentives, training, and many other activities all influence the behavior of personnel.

- **People achieve high levels of performance based on expectations and feedback.** People are influenced by incentives. If there is a reward for performing a task a certain way, people tend to perform it that way more often. If there is a punishment for the behavior, people generally do less of it. By rewarding personnel for desired behaviors and punishing personnel for undesirable behaviors, organizations can influence their behavior. Rewards and punishments should be focused on behaviors, not outcomes. Rewarding outcomes (i.e., getting the job done) can lead to personnel accomplishing the goal (task) using undesirable methods such as shortcuts and workarounds. Every time an employee completes a task in undesired ways without consequences, the employee builds confidence in his or her ability to "get away with it" again. Proper behavior should generate the desired outcome most of the time. Achieving the proper outcome using undesirable methods should be strongly and consistently discouraged. Many rewards and punishments are nonmonetary. Receiving recognition for a good idea, being able to select the tasks one performs, and being assigned to desirable or undesirable shifts or tasks are just some examples.
- **Personnel who are committed to their jobs perform better.** Personnel who are committed to their jobs feel an obligation to perform their best. People tend to be committed to doing their best when they (1) clearly understand the core values and performance goals, (2) have influence over what they do, (3) have the competencies to perform the jobs that are expected of them, and (4) are appreciated for their performance. Therefore, in addition to clear goals, rewards, and punishments, personnel must also have some control over their work. This can be accomplished by having personnel involved in the systems that affect their jobs, such as having operators involved in equipment design processes and procedural development.

> **Key Human Behavior Issues**
>
> - People are fallible, and even the best make errors.
> - Error-likely situations are predictable, manageable, and preventable.
> - Organizational processes and values influence individuals.
> - People achieve high levels of performance based on expectations and feedback.
> - Personnel who are committed to their jobs perform better.
> - Incidents can be avoided by understanding the reasons why human errors occur.

- **Incidents can be avoided by understanding the reasons why human errors occur.** If organizations understand the underlying reasons why human errors occur, they can set up effective systems for managing them. Systematic analyses of human behavior before incidents occur (proactive analyses) and systematic analyses of incidents themselves (reactive analyses) provide insight on how to design systems for successful human performance.

With these principles in mind, a COO/OD system seeks to eliminate the effects of human errors through the implementation of management systems that prevent, detect, and correct human errors.

4.3 WHAT IS A HUMAN ERROR?

Human error is any human action (or lack thereof) that exceeds some limit of acceptability (i.e., an out-of-tolerance action) where the limits of human performance are defined by the system. As used herein, the term "human error" includes slips, lapses, and intentional violations that may contribute to or result in accidents.[5]

> Human errors are gaps or differences between acceptable and actual behavior or performance.

Human errors are personnel performance gaps or differences between acceptable and actual performance by anyone at any level of the organization. Humans may fail to perform a required task (errors of omission) or perform a required task incorrectly (errors of commission). The definition of human error only makes sense when the limits of acceptability (or acceptable performance) are specified. In some cases, acceptable performance is specified by the order of steps in a procedure. For example, to start up a unit, the steps must be performed in the order specified by the procedure. If the procedure does not specify the order of some actions, then the worker performs the task correctly regardless of the order in which the actions are performed. Likewise, human errors are defined in terms of compliance with standards (acceptable performance). If the procedure specifies that the flow rate should be set at 45 to 55 gpm, then it would be a human error to set the flow rate to any value outside the specified range.

The same action can be a human error at one facility and acceptable behavior at another, depending upon the acceptable performance set by each facility. For example, at one facility, personnel are instructed to tighten packing on leaking valves without reporting or logging them. At another facility, there is a requirement to report and log each packing leak. Therefore, if an operator does not report the packing leak, it is either acceptable performance or a human error, depending upon the established facility standards.

Table 4.1 provides some examples of typical human errors that might be identified at a facility using the concept of differences between acceptable and actual performance to define what is meant by an error. Most of these human errors do not result in immediate consequences; however, they are still considered human errors.

---

[5] Current process safety-related definitions can also be found on the CCPS Web site.

TABLE 4.1. Examples of Personnel Performance Gaps (Human Errors)

Situation	Acceptable Performance	Actual Performance (Error)
Preparing to work on a valve	▪ Operations and maintenance personnel work together to prepare the valve for the work ▪ Maintenance communicates to the operations group the work that will be done ▪ Operations puts the process in a safe condition ▪ Maintenance verifies that the equipment is properly isolated	▪ Maintenance does not always verify proper isolation of the equipment because they rarely find problems and sometimes they are pressured to complete the job quickly
Shift turnovers	▪ Operators from both shifts (on-coming and out-going) discuss the work performed on the previous shift, equipment out of service, plans for the on-coming shift, and panel indications ▪ The turnover lasts long enough that the on-coming operators are comfortable with their understanding of the facility status and planned activities	▪ Shift turnover gets shortened because one operator arrives late or the other needs to depart the facility "on time"
Switching operating equipment	▪ The operator opens and closes specified valves in the sequence dictated by the operating procedure	▪ The operator manipulates the valves in a more convenient, but incorrect, sequence to save a few steps
Filling a tank	▪ The operator is supposed to set the fill rate at 45 to 55 gpm	▪ The operator sets the fill rate at 75 gpm to complete the job faster
Reviewing a new operating procedure	▪ Once the procedure is written, it is reviewed for technical issues by two operators: an operator normally assigned to the task and an operator who may be assigned to the task when the facility is operating short-handed	▪ The normally assigned operator signs off on the procedure without reviewing it, believing that the other operator will review the procedure in detail
Designing a new control system	▪ The design of the process requires multiple power sources to meet reliability requirements	▪ The designer eliminates one of the backup power supplies to meet budget constraints without getting approval for changing the reliability target

In all the cases outlined in the table above, there is a gap between acceptable and actual performance. In some cases, the performance gap might contribute to an immediate negative consequence for the facility. In other cases, additional safeguards prevent the performance gap from contributing to a loss. A COO/OD system tries to eliminate all deviations from acceptable performance, regardless of whether they have the potential for immediate negative consequences.

A COO/OD system is not trying to control every aspect of human performance; however, if there is a specified method or way to perform the task, then the goal of the COO/OD system is to have the task performed in accordance with the specified method every time. For tasks that are not important to the acceptable outcome, no detailed methods may be specified and workers may perform the tasks in whatever manner they choose. The goal of a COO/OD system is to drive out all deviations in performance.

The goal of the COO/OD system is to drive out all deviations in performance.

There are two general approaches to eliminating the performance gaps: change the actual behavior or change the process or system so that it is more tolerant of variation. COO/OD focuses on the first approach: the organization implements processes to reduce the frequency of the performance gap by changing the way people behave. However, given human limits and imperfections, the organization should also look for ways to reduce the consequences associated with the error by (1) making the system more tolerant of variation, (2) eliminating or relaxing the requirements, or (3) implementing other safeguards.

4.4 COMMON MISCONCEPTIONS ABOUT HUMAN PERFORMANCE

- **Punishing the people who make mistakes eliminates the mistakes.** Blaming people will temporarily reduce errors only in situations where fear of punishment is the primary driver behind personnel performance. However, this is just one of many drivers behind human performance. If the organization sets personnel up for failure and then blames individuals when they fail to perform well, punishment usually results in worse performance as personnel become resentful.
- **Training is the solution to all human performance problems.** Most human errors are not the result of a lack of knowledge or lack of skill (two areas where training can help). In many cases, errors are the product of a poorly designed human-machine interface (poor labeling, poor lighting, poor layout, etc.). Sometimes, people know what to do and how to do it, but they choose to do it differently. Other management systems, such as COO/OD, must be used to influence the behaviors.
- **Reward the right outcomes and everyone will behave properly.** Rewarding the desired outcomes encourages achievement of the goal by whatever means necessary. Shortcuts will often be used to achieve high production rates if the reward is based only on the outcome and there is no punishment for achieving the goal using undesirable methods. A better

approach is to reward and reinforce the use of the right process. Reinforcing the desired behaviors should result in the right outcomes.

- **Experienced personnel do not make errors.** Experienced personnel *do* err. The overall frequency of errors may be lower, and they may be more likely to detect and correct their errors before they have an adverse outcome. However, experienced personnel are more likely to make some errors of omission precisely because they are so familiar with routine tasks. Experienced workers who are bored or complacent are also more likely to make errors.
- **All errors must be eliminated.** It would be nice to eliminate all human errors. However, setting this as a goal or believing that this is possible leads to poor use of the organization's resources. When errors will have unacceptable consequences, methods for preventing the error, as well as detecting and correcting the effects of the errors, must be put in place to reduce the associated risk to a tolerable level. Organizations should not expend undue effort addressing performance variations with tolerable risk.
- **If everyone is held accountable, they will do the right thing.** Accountability is an important aspect of human performance management, and it is a tenet of COO/OD. No management system can function without a degree of personal accountability. However, holding personnel accountable for problems in the management systems they cannot control does not reduce human errors; it simply causes frustration, resentment, and poor performance.

Common Misconceptions

- Punishing the people who make mistakes eliminates the mistakes.
- Training is the solution to all human performance problems.
- Reward the right outcomes and everyone will behave properly.
- Experienced personnel do not make mistakes.
- All errors must be eliminated.
- If everyone is held accountable, they will do the right thing.

4.5 CATEGORIES OF HUMAN ERRORS

Human errors can be categorized in a number of different ways. The goal in identifying these categories is to identify solutions to prevent, detect, and/or correct the human errors.

The skill-, rule-, knowledge-based (SRK) approach refers to the degree of conscious control exercised by the individual over his or her activities. Figure 4.2 shows the continuum between conscious and automatic behavior in these three categories. The SRK approach is a widely used classification of industrial tasks that was developed by Jens Rasmussen and expanded upon by James Reason (Refs. 4.3, 4.4, and 4.5).

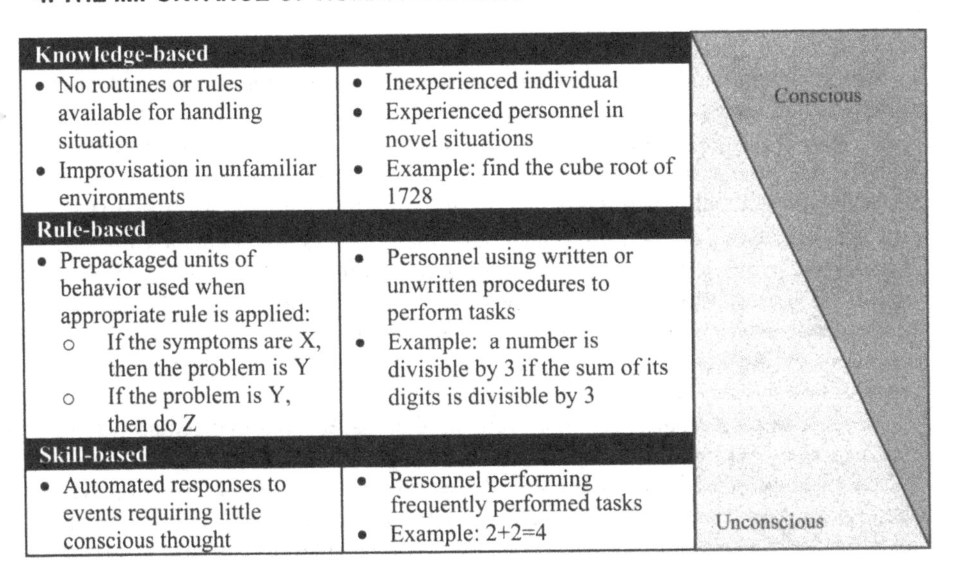

Knowledge-based	
• No routines or rules available for handling situation • Improvisation in unfamiliar environments	• Inexperienced individual • Experienced personnel in novel situations • Example: find the cube root of 1728
Rule-based	
• Prepackaged units of behavior used when appropriate rule is applied: o If the symptoms are X, then the problem is Y o If the problem is Y, then do Z	• Personnel using written or unwritten procedures to perform tasks • Example: a number is divisible by 3 if the sum of its digits is divisible by 3
Skill-based	
• Automated responses to events requiring little conscious thought	• Personnel performing frequently performed tasks • Example: 2+2=4

FIGURE 4.2. The Continuum Between Conscious and Automatic Behavior (Ref. 4.5)

- **Knowledge-based behaviors.** Humans carry out tasks in an almost completely conscious manner. This would occur in a situation where a beginner was performing the task or where an experienced individual was faced with a completely novel situation. In either case, the individual would have to exert considerable mental effort to assess the situation, and his or her responses are likely to be slow. In addition, when using knowledge-based behaviors, personnel monitor the system responses to their actions to determine whether the desired outcome occurred. This also slows their responses.
- **Rule-based behaviors.** These behaviors generally involve the execution of some set of predefined rules that the individual believes are appropriate for the situation. Because the desired actions (i.e., rules) have already been defined (e.g., through a written or unwritten procedure), the individual tends to perform these actions with little thought. Rule-based behaviors have a level of engagement between skill- and knowledge-based behaviors.
- **Skill-based behaviors.** This mode refers to the smooth, almost automatic execution of highly practiced, largely physical actions in which there is virtually no conscious monitoring. Skill-based responses are generally initiated by some specific event. For example, the requirement to operate a valve, which may arise from an alarm, a procedure, or another individual, will result in the highly practiced operation of opening the valve being executed largely without thought. Without any additional safeguards, skill-based errors are difficult to prevent because of the lack of conscious engagement by the performer.

The above classifications for human error provide a framework for analyzing variations in human behavior within a task. Potential safeguards can then be formulated to address each error type, as shown in Table 4.2.

Figure 4.3 shows a hierarchy of controls against potential human errors. In general, the less the implementation of the safeguard relies on the end user[6], the more reliable the safeguard tends to be. For example, passive safeguards (using less reactive, toxic, or flammable materials; the size of a containment dike; the thickness of a pipe; the pressure rating of a vessel) tend to be more reliable than active safeguards that require routine testing and maintenance. Testing and maintenance activities may be required, even for passive systems, to maintain the reliability of the safeguards. However, these activities are also opportunities to introduce new failure modes into the process.

Engineered safeguards are more reliable than safeguards that require prompt human actions. However, as dependence on the end user is decreased, the potential for failure is decreased but not eliminated. Instead, it is just moved to other parts of the organization, as shown in Figure 4.4. For example, operators cannot do much to influence the reliability of a passive safety system like the spacing of vessels. The effectiveness of spacing is largely determined by how it was designed and constructed. However, engineers must correctly calculate the thermal radiation flux and/or blast loads the spacing is intended to mitigate. As another example, consider an organization that had difficulties obtaining and analyzing liquid chemical samples using a manual sampling process. They decided to install an automated sampling system to address the problem. This system largely eliminates the need for the operators and laboratory technicians to perform ongoing sampling and analysis work. However, different errors can now be introduced into the design of the hardware and software for the automatic sampling system. These errors will likely be detected and corrected before consequences occur because (1) there is usually a large time lag between the introduction of the error and the consequences and (2) multiple design and hazard review steps are normally built into the engineering and construction processes.

> Engineers should actively solicit input and feedback from end users when designing safeguards against human error.

Designers and engineers are no more reliable in their individual performance than end users. However, when people have time to think through a situation and test a few solutions (like engineers usually can) and the consequences are not immediate, they generally come up with better solutions than people who have limited time for considering and evaluating multiple alternatives (like operators and mechanics.

---

[6] The end user is the person who uses the system. For most process industry facilities, these are operations and maintenance personnel. For a consumer product, it would be the person who uses the product at home. For a spare parts management database, it would be the stores personnel, purchasers, and maintenance personnel.

TABLE 4.2. Examples of Potential Safeguards for SRK Error Types

Error Type	Potential Safeguards
Knowledge-based	Knowledge-based training Decision-support tools Crew-resource management Traditional rewards and incentives
Rule-based	Effective written procedures Independent verification Information at the point of use
Skill-based	Skill-based training Error-proofing (also known as Poke-Yoke) Interlocks Ergonomics

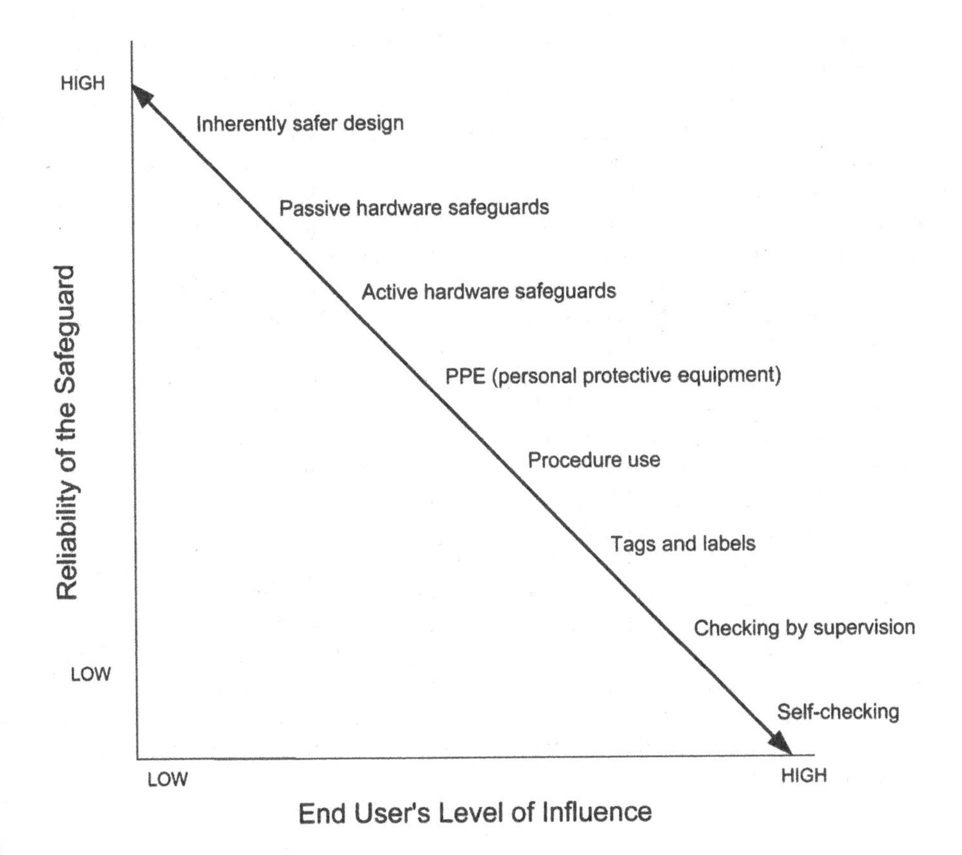

FIGURE 4.3. Reliability of Safeguards Versus Reliance on the End User

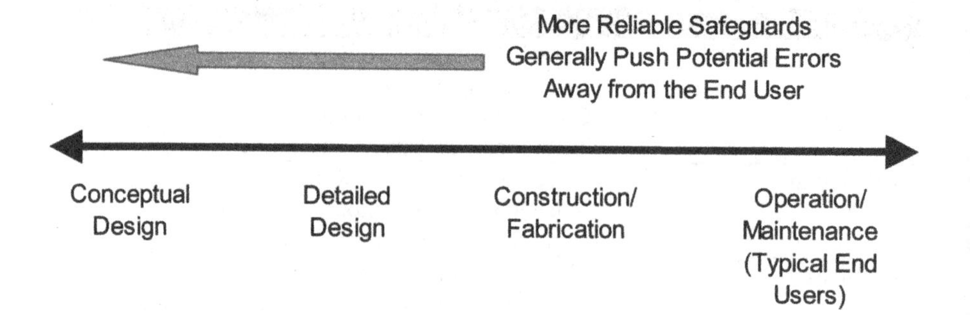

FIGURE 4.4. More Reliable Safeguards Push Errors Away from End Users

Another factor to consider when implementing hardware- and software-based safeguards is that they often replace, instead of supplement, existing safeguards. For example, a high-level switch can be installed to close the inlet valve to a tank automatically to prevent overflows. The facility may still require the operators to stay by the tank as it fills, believing that the automatic system is a backup in case the operator fails to close the valve. However, if the facility does not enforce the requirement and encourages personnel to fill multiple tanks simultaneously, operators will rely upon the automatic system and no longer monitor the tank fills. Thus, the risk of a future spill may not be reduced as expected.

Project managers should choose the lowest cost option that reduces risk to a tolerable level across a project's life cycle. When project managers try to minimize initial capital costs, they often choose a less effective safeguard that relies more heavily on the end user. However, if life-cycle costs are properly evaluated, it is usually more advantageous to select inherently safer approaches or provide engineered controls in the original design than to retrofit the system later. A risk-informed decision would select the best means for reliably accomplishing the facility's long-term goals. The field of resilience engineering offers other ways to incorporate human and organizational risk in life-cycle analyses.

CCPS's *Concept* book entitled *Human Factors Methods for Improving Performance in the Process Industries* (Ref. 4.6) includes a human factors toolkit that describes twenty-eight human factors topics that can be used to address human performance issues. These topics include equipment design issues, labeling, shift work issues, procedures, management of change, and competence management. CCPS has also published *Guidelines for Preventing Human Error in Process Safety* (Ref. 4.7). It provides further details on developing and managing a human factors program.

4.6 HUMAN ERROR INITIATORS

When and where should you be on the lookout for human errors? There are many types of error-likely situations or conditions in which human errors are more likely to occur. Table 4.3 provides three examples of error-likely situations. Additional examples of error-likely situations can be found in the online material that accompanies this book.

4.7 HOW DOES A COO/OD SYSTEM PREVENT AND MITIGATE HUMAN ERRORS?

The goal of the COO/OD system is to continuously search for error-likely situations and then take actions to prevent, detect, and correct any actual or potential errors identified. This helps organizations achieve the goal of zero incidents. Remember, the organization is trying to achieve zero incidents, not zero human errors. We know that human errors will occur no matter how hard we try to eliminate them. However, with the proper safeguards to prevent, detect, and correct the human errors, a goal of zero incidents can be achieved.

For each of the error-likely situations shown in Table 4.3 and in the accompanying online material, COO/OD system elements can help address the situation. Further details on the methods to address similar situations can be found in Chapters 3, 5, 6, and 7 of this book.

4.8 RELATIONSHIP BETWEEN COO/OD AND OTHER COMMON HUMAN PERFORMANCE TOOLS

This section looks at the relationship between COO/OD systems and other human performance tools that an organization might already be using. Historically, these tools have primarily focused on personal safety issues. However, they can also be applied to human behaviors affecting process safety. These tools include:

- Behavior-based (BB) programs
- Antecedent-behavior-consequence (ABC) programs
- Human performance technology (HPT) approach

All three of these approaches recognize the influence that appropriate consequences, both positive and negative, have on human performance. There are three dimensions of consequences that are usually considered as part of these approaches: positive versus negative, immediate versus delayed, and certain versus uncertain. Positive, immediate, certain (P-I-C) feedback is the most influential on behavior, while negative, delayed, uncertain (N-D-U) feedback is the least influential.

In a typical situation, multiple types of consequences are present. For example, consider a mechanic repairing a cyclone separator. The procedures require him to wear a hard hat, respirator, safety goggles, gloves, and steel-toed boots during the activity. The mechanic finds that wearing the steel-toed boots is not much of an inconvenience (not a significant enough negative, immediate, certain consequence),

TABLE 4.3. Examples of Error-Likely Situations

Error-Likely Situation	Anticipated Performance Gaps	How to Address the Situation
Task Issues		
Simultaneous, multiple tasks	Workers will attempt to multitask, but they will inevitably focus on one activity to the detriment of the other (e.g., talking on a mobile phone while driving). The worker may confuse the steps in these tasks or attempt to apply the procedure for one task to the other task.	Provide error-proofing to prevent misapplication of the procedure for one task to another task. Concentrate efforts on tasks that are similar but have a few critical steps that are different. For safety-critical tasks such as driving, consider limiting simultaneous performance of other distracting tasks such as use of a mobile phone.
Work Environment		
Unreliable/inoperable equipment	When equipment does not work the way it should, personnel "invent" new ways to overcome the difficulty and perform the task. Usually these innovative solutions bypass the MOC review process.	Maintain equipment in operable condition and promptly perform repairs to return equipment to service. Use the MOC process to control temporary and permanent changes.
Individual Capabilities		
Lack of a sufficient mental model for a task	Mental models are humans' simple models for complex processes. An inaccurate model often results in mistakes. If an individual's (or group's) model of the relationship between temperatures, pressures, flows, and levels throughout the process is wrong, they will not be able to respond appropriately to novel situations.	Provide personnel with sufficient training and experience to understand the concepts behind the operation of the equipment. This will allow them to correctly diagnose the causes of abnormal operations. Use mock drills and simulations to allow personnel to explore process behavior beyond its normal limits. Emphasize why the tasks are structured the way they are, not just how to perform the task.

and he has seen heavy objects fall during the activity (he could avoid a significant negative, immediate, but uncertain [it is unlikely to happen to him] consequence [foot injury]). Therefore, he wears the boots. The respirator is a different story. He cannot see why he needs to wear it. If he does inhale the dust, the health effects (negative) may affect him (uncertain) years later (delayed). No one is likely to punish him for not wearing the respirator (a negative, immediate, but highly uncertain consequence). Not wearing the respirator provides greater comfort (a P-I-C consequence). It is unlikely that anyone will comment positively if he does wear the respirator (a positive, immediate, uncertain consequence). As a result, the mechanic does not wear the respirator.

> The performer, not the person delivering the consequence, determines whether a consequence is positive or negative.

Because there are many different consequences (positive/negative, immediate/delayed, and certain/uncertain), people will usually weigh the positive and negative consequences of the behavior to decide on a course of action. The course of action where the positive consequences outweigh the negative consequences is usually what will be implemented, regardless of the methods stated by the organization. The COO/OD system seeks to make the desired performance the most attractive way to perform the task; in other words, to make the right way the way that personnel will want to choose.

4.8.1 Behavior-Based Programs

One structured method used by many organizations to provide for the ongoing involvement of front-line personnel in the COO/OD system is the BB program (Refs. 4.6 and 4.8). Figure 4.5 shows the basic process flow diagram for a BB program. BB programs provide ongoing (immediate) feedback/consequences for front-line personnel regarding their behaviors. As noted before, both COO/OD systems and human factors programs focus on influencing the behavior of personnel, not the outcomes. There are two underlying premises of BB approaches:

1. If personnel exhibit the right behaviors, they are more likely to generate the correct outcomes.
2. P-I-C consequences are more likely to generate the proper behaviors than other types of feedback.

"Consequences" as used here is a very broad term. It includes positive and negative consequences, such as recognition, rewards, punishments, verbal comments, work assignments, time savings, embarrassment, and routine reinforcement.

The BB process is intended to involve all front-line personnel in (1) providing feedback on behaviors of front-line personnel and (2) identifying performance gaps. The program is centered on the same activities as the COO/OD system: the performance of front-line personnel.

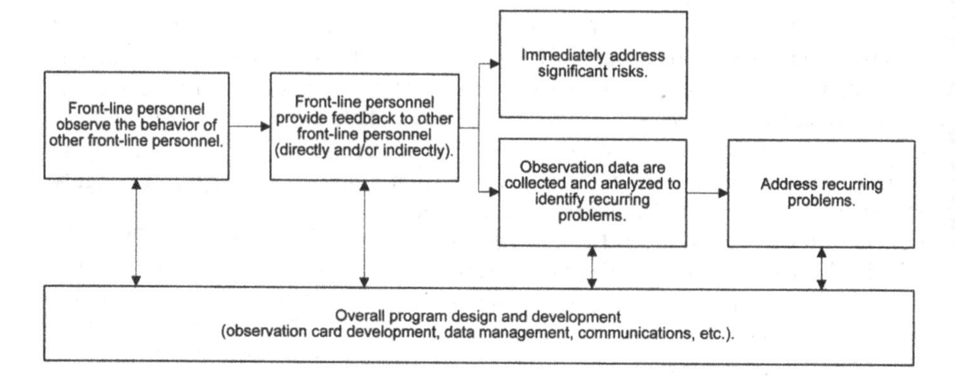

FIGURE 4.5. Behavior-Based Program Flowchart

If a facility already has a BB program in place, a COO/OD system can improve its performance. The COO/OD process will clarify and stress the importance of the desired behaviors expected of personnel, and the BB program will reinforce those actions through the positive and negative observations made by personnel and the associated immediate, certain outcomes.

Another advantage of BB programs is that they focus on leading indicators of performance. Some indicators, like incident rates and downtime, are lagging indicators. Lagging indicators are generally easy to measure, but they only identify deteriorated performance after losses have occurred. Leading indicators, such as housekeeping, the number of outstanding work orders, and the number of out-of-date procedures, are generally more difficult to measure, but they are usually better predictors of future performance. BB programs focus on leading indicators by observing front-line personnel behaviors. When undesirable behaviors are observed, actions are taken to eliminate the performance gaps before an incident with significant consequences occurs.

4.8.2 Antecedent-Behavior-Consequence Programs

Another approach to understanding human behavior is the ABC analysis method (Ref. 4.9). Figure 4.6 shows the basic approach used for ABC analysis. The antecedents (the conditions that exist prior to the behavior) and the consequences (what happens from the perspective of the person who performs the task) influence the behaviors of personnel. If performance gaps are identified, then the antecedents and/or consequences are adjusted to achieve the desired behaviors.

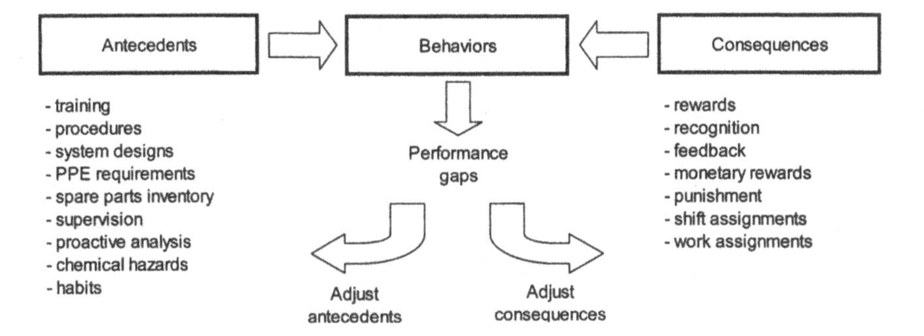

FIGURE 4.6. Antecedent-Behavior-Consequence Analysis Flowchart

ABC analysis can also provide leading indicators of performance when it is applied to at-risk behaviors or near-miss incidents in an attempt to modify behaviors before a serious incident occurs. It can also be used to understand human performance after an incident has occurred as part of a root cause analysis program. If a facility already has an ABC program in place, a COO/OD system can enhance its performance by:

- Clearly identifying the desired behaviors
- Improving the antecedents (management systems such as COO)
- Providing appropriate consequences (systems such as OD that provide positive/immediate/certain feedback for desirable behaviors and negative/immediate/certain feedback for undesirable behaviors)

4.8.3 Human Performance Technology Approach

The HPT approach (Ref. 4.10) evaluates human behavior in a systematic manner. It is structured around a variation of the Plan-Do-Check-Adjust approach. A flowchart for the HPT approach is shown in Figure 4.7

The basic steps of HPT include the following[7]:

1. Identify performance gaps
2. Analyze causes
3. Select, design, and develop prevention, detection, and correction safeguards
4. Implement the safeguards
5. Evaluate

[7] Additional details on the HPT approach can be found in the chapter references.

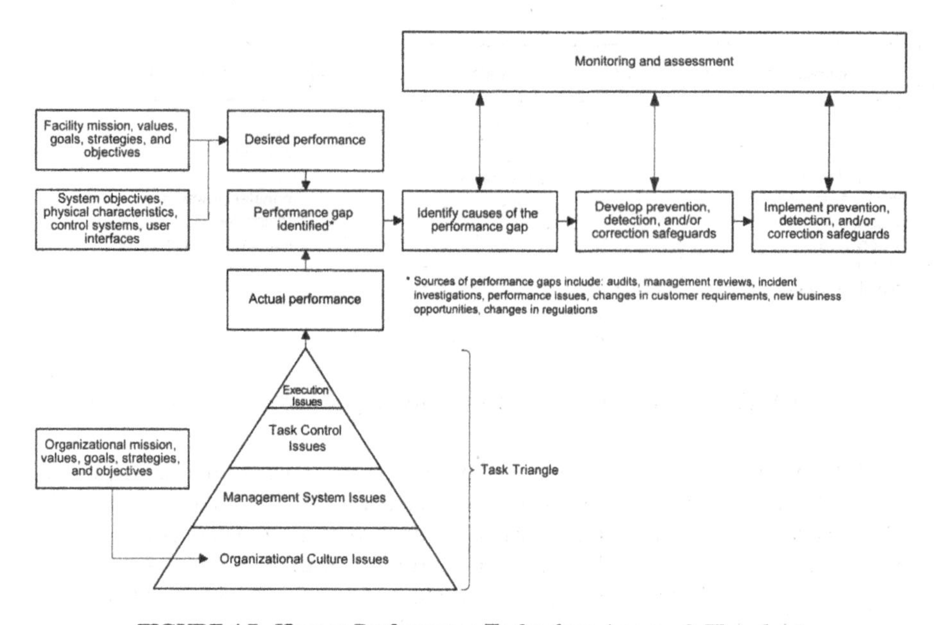

FIGURE 4.7. Human Performance Technology Approach Flowchart

These steps can be applied in a very formal manner, such as a structured audit or formal incident investigation. Support organization personnel usually lead the formal application of these steps. However, the prevention, detection, and correction processes implemented as a result often include actions implemented by front-line personnel. Safeguards that result from a more formal application of the approach are usually more broadly focused and tend to affect the lower portions of the task triangle.

Informal applications of the approach are often led and implemented by front-line personnel with little involvement by support organizations. Safeguards that result from a more informal application of the process are usually more narrowly focused and tend to affect the top portion of the triangle.

If your organization is using the HPT approach, a COO/OD system can enhance its performance through improvements in the organizational management systems and emphasis on the identification of performance gaps.

4.9 GETTING EVERYONE INVOLVED IN HUMAN FACTORS

One basic aspect of both an effective COO/OD system and an effective human factors program is the involvement of all the facility staff. Every group in the organization has some role in the identification of performance gaps and the design, development, and implementation of safeguards. Table 4.4 identifies a few key activities in an effective human factors/COO/OD system.

TABLE 4.4. Typical Human Factors/COO/OD Activities by Group

Group	Typical Human Factors/COO/OD Activities
Managers	• Monitoring key human factors/COO/OD metrics • Reinforcing positive behaviors using P-I-C feedback
Supervisors	• Consistently enforcing the rules • Providing P-I-C rewards for desirable behaviors
Engineering	• Considering end-user needs in the design of processes • Addressing the end-users' issues with existing processes • Incorporating human factors into the design of processes
Research and Development	• Investigating inherently safer alternatives • Incorporating human factors principles into original designs • Soliciting input from end users of the equipment at the early design stages
Operations	• Identifying error-likely situations • Reinforcing the use of good human factors tools • Participating in the design of equipment and procedures • Identifying potential problems from an operations perspective
Maintenance	• Correcting chronic equipment performance issues • Reinforcing the use of good human factors tools • Participating in the design of equipment and procedures • Identifying potential problems from a maintenance perspective
Warehouse/Stores	• Maintaining human factors aids (e.g., aisle markers, labeling, visual reorder level indictors, part number labeling, and repair kits) used in the warehouse • Identifying human factors improvements applicable to the warehouse
Procurement	• Addressing human factors issues in the procurement specifications for spares and equipment
Training	• Incorporating human factors issues and principles into training for personnel • Identifying performance issues that can be addressed through human factors tools
Procedure Writers	• Consulting end users on the development of procedures • Developing procedures consistent with human factors principles

4.10 HUMAN FACTORS METRICS

Metrics can be used to monitor the health of the organization's human factors efforts. Metrics can be divided into leading and lagging indicators. Leading indicators are predictors of future performance. Lagging indicators monitor the current and past performance of the facility to identify areas of significant risk.

For the human factors program, examples of leading indicators include the following:

- Number of human-machine interface deficiencies and near-miss reports – an indication of the priority assigned to human factors issues
- Number of nuisance and always-on alarms – a measure of the normalization of deviant conditions
- Average time to complete repairs on the human-machine interface – an indication of the priority assigned to resolving human factors issues
- Number of inspection deficiencies related to lighting and labeling – an indication of the priority assigned to resolving human factors issues
- Housekeeping status – to assess the cleanliness of the facility
- Number of observations made as part of a BB program – a measure of the level of engagement in the process by front-line personnel
- Number of ABC analyses performed – an indication of the health of the program
- Number of modifications made to processes, procedures, or practices in reaction to human factors concerns – a measure of the breadth and level of human factors involvement within the facility
- Level of resources (time or money) devoted to human factors improvement programs – a measure of management commitment to resolve human factors issues

Examples of lagging indicators include the following:

- Injury rates – many injuries involve human factors issues
- Downtime – many unplanned shutdowns or outages involve human factors issues
- Quality deficiencies – many quality deficiencies involve human factors issues
- Unplanned actuations of safety systems – many are caused by human factors issues
- Incidents – many have human factors issues identified as causes

The performance metrics for the COO and OD elements in Chapter 7 can also reflect the effectiveness of the overall facility human factors program. In addition, the CCPS, the API, and the U.K.'s Health and Safety Executive have published guidelines specifically for process safety metrics (Refs. 4.11, 4.12, 4.13, and 4.14).

Not all of the metrics listed will be feasible and appropriate for a facility. After reviewing the metrics and the resources needed to collect and analyze the data, the facility and organization should select appropriate metrics to implement.

4.11 SUMMARY

This chapter has examined the role of human factors in the development and implementation of a COO/OD system. By understanding the factors that influence human behavior, the effectiveness of the COO/OD system can be improved.

4.12 REFERENCES

4.1. U.S. Nuclear Regulatory Commission, "Backgrounder on Chernobyl Nuclear Power Plant Accident," Washington, D.C., April 2009.

4.2. Atherton, John, and Frederic Gil, *Incidents That Define Process Safety*, Center for Chemical Process Safety of the American Institute of Chemical Engineers, John Wiley & Sons, Inc., Hoboken, New Jersey, 2008.

4.3. Rasmussen, Jens, *Information Processing and Human-Machine Interaction: An Approach to Cognitive Engineering*, Elsevier Science Ltd, Amsterdam, The Netherlands, 1986.

4.4. Rasmussen, J., "Human Errors: A Taxonomy for Describing Human Malfunction in Industrial Installations," *Journal of Occupational Accidents*, ScienceDirect/Elsevier, Amsterdam, The Netherlands, Vol. 4, 1982, pp. 311-333.

4.5. Reason, James, *Human Error*, Cambridge University Press, Cambridge, England, 1990.

4.6. Center for Chemical Process Safety of the American Institute of Chemical Engineers, *Human Factors Methods for Improving Performance in the Process Industries*, John Wiley & Sons, Inc., Hoboken, New Jersey, 2007.

4.7. Center for Chemical Process Safety of the American Institute of Chemical Engineers, *Guidelines for Preventing Human Error in Process Safety*, John Wiley & Sons, Inc., Hoboken, New Jersey, 2004.

4.8. Krause, Thomas R., *The Behavior-Based Safety Process: Managing Involvement for an Injury-Free Culture, 2nd Edition*, John Wiley & Sons, Inc., Hoboken, New Jersey, 1996.

4.9. McSween, Terry E., *The Values-Based Safety Process: Improving Your Safety Culture with Behavior-Based Safety, Second Edition*, John Wiley & Sons, Inc., Hoboken, New Jersey, 2003.

4.10. Van Tiem, Darlene M., James L. Moseley, and Joan Conway Dessinger, *Fundamentals of Performance Technology: A Guide to Improving People, Process, and Performance, Second Edition*, International Society for Performance Improvement, Silver Spring, Maryland, 2004.

4.11. Center for Chemical Process Safety of the American Institute of Chemical Engineers, *Guidelines for Process Safety Metrics*, John Wiley & Sons, Inc., Hoboken, New Jersey, 2009.

4.12. ANSI/API Recommended Practice 754, *Process Safety Performance Indicators for the Refining and Petrochemical Industries*, American Petroleum Institute, Washington, D.C., April 2010.

4.13. U.K. Health and Safety Executive, *A Guide to Measuring Health & Safety Performance,* London, England, December 2001.

4.14. U.K. Health and Safety Executive, *Developing Process Safety Indicators: A Step-by-Step Guide for Chemical and Major Hazard Industries*, HSE Books, London, England, 2006.

4.13 ADDITIONAL READING

- Perrow, Charles, *Normal Accidents: Living with High-Risk Technologies,* Princeton University Press, Princeton, New Jersey, 1999.
- Strauch, Barry, *Investigating Human Error: Incidents, Accidents, and Complex Systems*, Ashgate Publishing Limited, Aldershot, England, 2004.
- Dekker, Sidney, *The Field Guide to Understanding Human Error*, Ashgate Publishing Limited, Aldershot, England; 2006.
- Lipman-Blumen, Jean, and Harold J. Leavitt, *Hot Groups: Seeding Them, Feeding Them, & Using Them to Ignite Your Organization*, Oxford University Press, Inc., New York, New York, 1999.
- Buckingham, Marcus, and Curt Coffman, *First, Break All the Rules: What the World's Greatest Managers Do Differently*, Simon & Schuster, New York, New York, 1999.
- Hollnagel, E., D. Woods, and N. Leveson, organizers, *Proceedings of the First International Symposium on Resilience Engineering*, Söderköping, Sweden, October 25-29, 2004.
- U.K. Health and Safety Executive, *Reducing Error and Influencing Behaviour*, HSE Books, London, England, 1999.
- Reason, James, *The Human Contribution: Unsafe Acts, Accidents and Heroic Recoveries*, Ashgate Publishing Limited, Farnham, England, 2008.
- Hollnagel, Erik, *Barriers and Accident Prevention: Or How to Improve Safety by Understanding the Nature of Accidents Rather Than Finding Their Causes*, Ashgate Publishing Limited, Aldershot, England, 2004.
- Mager, Robert F., *Analyzing Performance Problems or You Really Oughta Wanna: How to Figure Out Why People Aren't Doing What They Should Be, and What to Do About It, Third Edition*, The Center for Effective Performance, Inc., Atlanta, Georgia, 1997.
- Reason, James, *Managing the Risks of Organizational Accidents*, Ashgate Publishing Limited, Aldershot, England, 1997.
- Kinlaw, Dennis C., *Coaching for Commitment: Interpersonal Strategies for Obtaining Superior Performance from Individuals and Teams, Second Edition*, Jossey-Bass/Pfeiffer, San Francisco, California, 1999.

5 KEY ATTRIBUTES OF CONDUCT OF OPERATIONS

5.1 INTRODUCTION

As defined in Section 1.4, conduct of operations is the embodiment of an organization's values and principles in management systems that are developed, implemented, and maintained to (1) structure operational tasks in a manner consistent with the organization's risk tolerance, (2) ensure that every task is performed deliberately and correctly, and (3) minimize variations in performance.

- COO is the management systems aspect of COO/OD
- COO sets up organizational methods and systems that will be used to influence individual behavior and improve process safety
- COO activities result in specifying how tasks (operational, maintenance, engineering, etc.) should be done
- A good COO system visibly demonstrates the organization's commitment to process safety

This chapter explores the key attributes that form the foundation of an effective COO system. Operational discipline, which focuses on task performance, is addressed in detail in Chapter 6.

At its core, the goal of a COO system is to affect what people do or fail to do. COO helps establish the conditions necessary to achieve highly reliable performance, which is reflected in (1) predictable, consistent, and proper actions by the entire organization, (2) capable and stable processes, and (3) reliable plant operation. Therefore, it is a set of attributes that help ensure that people make the right choices at the right times, and that the organization operates in a safe, predictable, and reliable manner.

COO applies to the entire organization. Traditionally, some think of COO as a means to ensure that front-line personnel, such as operators and maintenance employees, adhere to a set of rules. COO is much more than that. It encompasses people, processes, and plant equipment. As described in Chapter 3, it starts at the executive level, but it reaches to the shop floor. COO extends beyond the operational "line" to include support functions such as engineering, quality control, and product/process development. A truly effective COO program touches the entire organization.

COO should be appropriately scoped. The COO system starts by identifying the standards (including policies, procedures, and practices) that are most critical, and helps ensure compliance with those standards. For example, there would likely be several safeguards to help prevent a single human error from purchasing and installing process equipment that does not conform to specifications, whereas there is likely no specified means to ensure that noncritical items are fit for purpose.

Bhopal – Failure of Multiple Protection Layers

The events that led to a large release of methyl isocyanate in Bhopal, India, on the night of December 3, 1984, are well documented. Thousands died and more than 100,000 were injured; clearly it was the worst chemical plant disaster in history. Many events contributed to the release, and the immediate cause has never been determined with certainty; however, it is clear that several safety systems were not maintained, were ignored, or were unavailable. These include:

- High pressure and high temperature alarms that were initially ignored because they were known to be unreliable
- Chillers that had been shut down in an effort to cut costs
- A flare that had been out of service for several months prior to the incident due to maintenance issues
- A scrubber that was out of service at the time, and when it was restarted it was ineffective because the strength of the caustic had not been maintained within the proper range

As Trevor Kletz points out in his book *What Went Wrong? Case Histories of Process Plant Disasters* (Ref. 5.1), it is relatively easy to purchase safety systems. However, it is much more difficult to keep them in operating condition, particularly when failure does not immediately impact production or other metrics that are tracked on a daily basis. Once management stops taking an active interest, workers are likely to focus on other issues as well. The attributes of COO presented in this chapter are designed to help ensure that focus on process safety is not lost or diverted simply because there have not been any recent process safety incidents.

Before reading further, review Figure 5.1, the COO/OD improvement and implementation cycle used throughout this book. Many readers are interested in establishing a new COO system (entering the cycle from the 12 o'clock position). The next step is to establish goals for each COO or OD element for key stakeholders to review, understand, and hopefully embrace. As described in Chapter 3, the goals need to be realistic (i.e., based on improvements that can be delivered by COO/OD). The goals should also be measurable and provide some tangible benefits that will motivate the organization to spend time, allocate resources, or otherwise expend effort. To move forward, strong commitment and active management support are required.

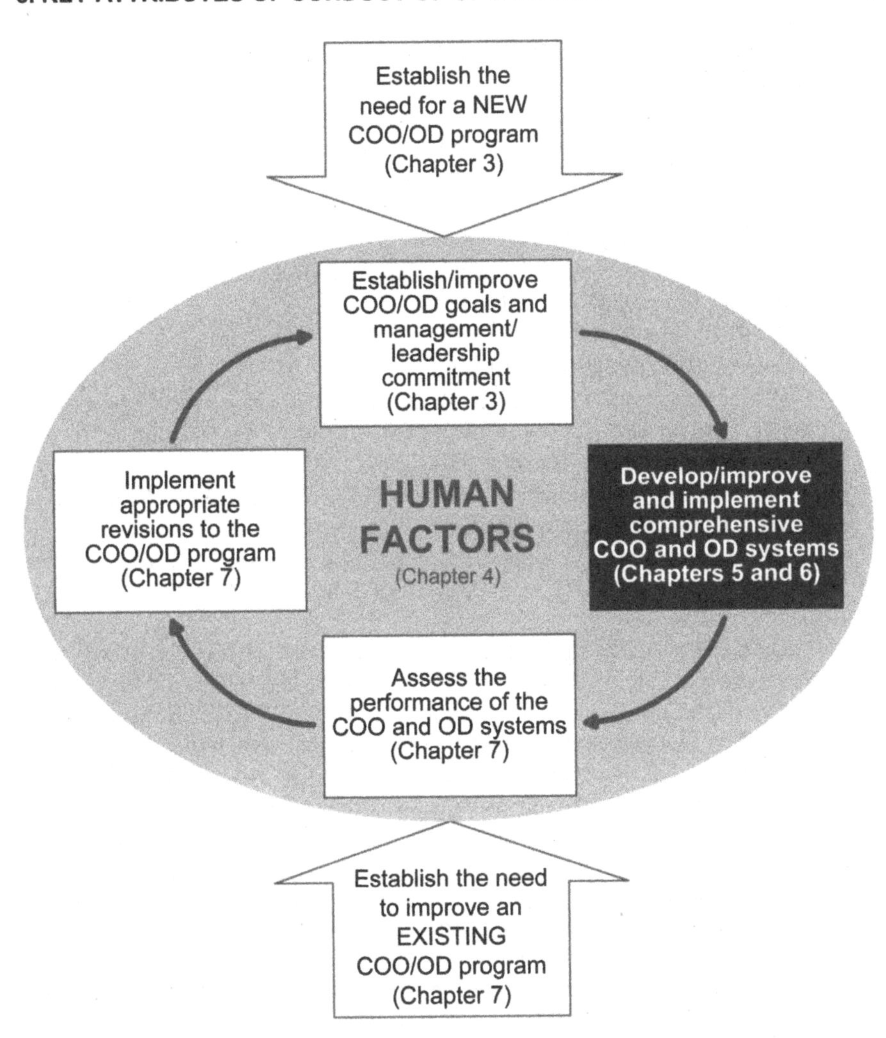

FIGURE 5.1. COO/OD Improvement and Implementation Cycle

This chapter, along with Chapter 6, addresses the box in the 3 o'clock position. Sections 5.4 through 5.7 describe attributes of an effective COO/OD system. Some, but not all, of the attributes are likely to be applicable to a given situation; readers should view these thirty attributes as a menu of ideas for implementing COO/OD in their organization. The narrative description of each attribute can be used to benchmark existing programs or identify desirable features for new programs. Once objectives are defined, the next step is to assess the current systems in place at a facility to identify which COO attributes are most likely to help the facility achieve the goals. Chapter 7 addresses this assessment process, along with ideas on how to implement a COO/OD system.

5.2 COO APPLIED TO PROCESS SAFETY MANAGEMENT SYSTEMS

The focus of this book is on COO as it applies to PSM systems. There are several published frameworks for PSM systems, including the following:

- CCPS RBPS management system (Ref. 5.2)
- ACC Responsible Care® Management System (Ref. 5.3)
- OSHA PSM regulation 29 CFR 1910.119
- EPA RMP rule 40 CFR 68
- Seveso II Directive (96/082/EEC)
- API Recommended Practice 75, *Recommended Practice for Development of a Safety and Environmental Management Program for Offshore Operations and Facilities* (Ref. 5.4)
- U.K. Statutory Instrument 2005 No. 3117, *The Offshore Installations (Safety Case) Regulations 2005*
- DOE Order 5480.19, Change 2, *Conduct of Operations Requirements for DOE Facilities* (Ref. 5.5), and other DOE orders

Although COO has been in place in various industries for decades, it was first proposed as a process safety element by the CCPS in the *Guidelines for Risk Based Process Safety* (Ref. 5.2). As shown in Figure 5.2, the CCPS's RBPS management system is based on four pillars, and COO is one of the elements in the Manage Risk pillar. It was included because other elements in that pillar will only be effective if there is a program that helps ensure reliable, consistent, and correct execution of the policies, procedures, and practices that make up the facility's risk management system.

COO supplements, rather than replaces, process safety and other EH&S management systems. This chapter does not focus on basic practices for operating and maintenance procedures, training, safe work practices, asset integrity, management of change, pre-startup safety review, and the like. Rather, it focuses on a program to help ensure the effectiveness of these and other PSM systems. An effective COO system supports PSM systems. At the other end of the scale, a nonexistent COO system can lead to a "check the box mentality," perhaps supporting or even institutionalizing ineffective practices.

Although COO is included in the Manage Risk pillar of CCPS's RBPS management system (the grouping of elements most closely associated with day-to-day facility operation), it spans all four pillars:

- As described in Chapter 3, COO/OD cannot be separated from several elements in the Commit to Process Safety pillar, such as process safety culture, compliance with standards, and workforce involvement.
- COO contributes to the Understanding Hazards and Risk pillar; it applies equally to the engineering office and the shop floor. It also applies to design reviews and hazard/risk analyses performed by third parties.

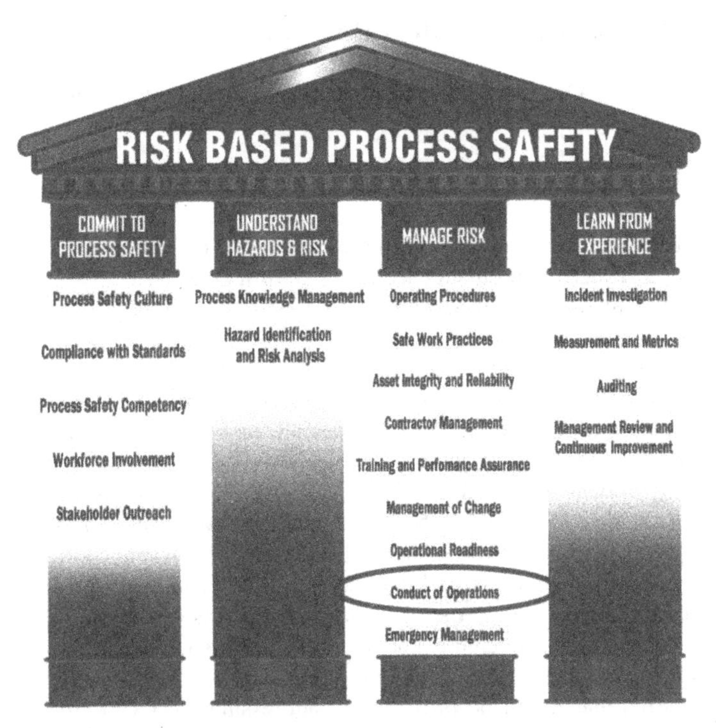

FIGURE 5.2. CCPS's Risk-Based Process Safety Management System (David Guss, Nexen Inc., 2008)

- An effective COO system demands that organizations Learn from Experience. The model presented in Chapter 3 involving inertia is true – a body at rest is clearly not moving forward.

COO is behavior-oriented, and organizations that strive to ensure that activities are conducted in a predetermined manner already have a COO system. This can feel a bit "fuzzy." However, the efficacy of a COO system can be measured in terms of overall organizational reliability. For example, answering the questions "How well do we follow specified procedures?" and "When we follow procedures do we achieve the intended result?" are tests of the effectiveness of a COO system. Thus, the inherent synergy between COO and EH&S and/or quality management systems can provide important feedback on the health of the COO system. While it is possible to exhibit effective COO/OD practices in one area, such as quality, and not expand these effective practices to other important aspects, such as process safety, this segmentation requires substantial effort to sustain. Effective COO/OD systems span the entire organization.

5.3 ORGANIZATION OF THIS CHAPTER

The attributes of COO presented in this chapter are divided into the following three groupings:

1. Foundational attributes that apply to all aspects of an operation (Section 5.4)
2. Attributes that apply to a single aspect of the operation, specifically people, process, and plant (Sections 5.5 through 5.7)
3. Management systems (Section 5.8)

Each COO attribute starts with a short summary followed by an example that helps demonstrate the attribute. In some cases, the examples are based on high-profile incidents. In other cases, the examples are based on events that were never known outside of the facility where they occurred. The stories are not intended to summarize significant historical accidents. The CCPS and others have done that elsewhere (Refs. 5.1, 5.6, 5.7, 5.8, and 5.9). Rather, readers should use each story as an example to help determine whether the attribute is relevant to their operation by asking questions such as "Could that happen here?" and "If it did happen here, would the consequences be tolerable?" Based on the answers to those and similar risk questions, readers are challenged to further evaluate the lessons from the example, to determine whether a similar gap exists in their organization, and to evaluate benefits that might be provided by addressing the COO attribute in that section.

There are limitless opportunities to improve COO. Consideration of the COO attributes presented in this chapter may expose more serious issues for some facilities than for others. Some facilities will already have in place a very sound program that addresses a particular attribute, or it will be clear that the attribute does not apply at all. In other cases, the reader will need to critically evaluate the attribute to determine whether it is likely to provide value for his or her facility. From this list of opportunities, the reader should prioritize the gaps and evaluate the effort required to implement a new system or improve an existing system. These decisions should be based on risk, relying on experience and sound judgment.

5.4 COO FOUNDATIONS

First and foremost, management must be committed to COO/OD and strive to embed effective COO/OD practices into the organization's culture. Nothing will undermine COO more quickly than management not "walking the talk." As the saying goes, "The best performance you should expect is the worst you personally demonstrate." The importance of leadership and management commitment is addressed in detail in Chapter 3.

Management commitment and leadership are only the first steps in developing an effective COO/OD system. Well-supported programs that are comprehensive and consistent lead to the effective COO system for people, process, and plant considerations that is described in Sections 5.5 through 5.7. However, the attributes in this section should be reviewed and understood prior to developing programs to address some or all of the COO attributes. They include the following:

1. Understand risk significance (and pay attention to what matters)
2. Establish standards that support the organization's mission and goals
3. Understand what can be directly controlled and what can only be influenced
4. Provide the resources and time necessary to complete tasks within standards
5. Ensure competency across the organization
6. Perform critiques and take corrective action

5.4.1 Understand Risk Significance

COO systems should focus people and systems on what really matters, not necessarily on what is easy to accomplish or measure. Without understanding the significance of risk, the COO system can deteriorate to the point where leaders enforce policies simply because they exist. The entire organization should understand what matters most and be able to link the standards that collectively make up the COO system to efforts to improve and sustain performance of risk-significant activities.

The 1986 loss of the space shuttle Challenger and its crew was due to o-ring failure. According to the *Report of the Presidential Commission on the Space Shuttle Challenger Accident* (Ref. 5.10):

> *The Space Shuttle's Solid Rocket Booster problem began with the faulty design of its joint and increased as both NASA and contractor management first failed to recognize it as a problem, then failed to fix it, and finally treated it as an acceptable flight risk.*

The joint was assigned Criticality 1 (the highest criticality, indicating that failure could lead to loss of life or loss of the space shuttle), but many believed that since the joint involved two o-rings, it could be downgraded to Criticality 1R (indicating that two, redundant systems would have to fail to cause a loss of life or loss of the vehicle). The commission's report goes on to note that:

> *The Problem Assessment System . . . still listed the [solid rocket booster] field joint as Criticality 1R on March 7, 1986, more than five weeks after the accident. . . . As a result, informed decision making by key managers [regarding the integrity of the joint and safety of the vehicle and crew] was impossible.*

Risk understanding starts with hazard identification. Once hazards are identified, a number of methods can be used to (1) analyze risk, (2) determine whether existing safeguards are adequate, and (3) recommend appropriate process changes, additional safeguards, or improvements to existing safeguards (Ref. 5.11). Safeguards, whether applicable to a single accident scenario (e.g., an interlock) or to the operation in general (e.g., the work permitting system), should be well understood, clearly documented, and enforced. Safeguards that significantly impact risk should be included in the COO system.

Facilities that understand risk significance:

- Perform balanced hazard identification and risk analysis studies that thoughtfully examine risk associated with high-, medium-, and low-frequency events
- Have a system in place to periodically reexamine and update hazard identification and risk analysis studies over the life cycle of the process/operation
- Document safeguards that are identified through various hazard evaluations and risk analyses
- Segment safeguards according to those that must always be in place, those that may be waived with special management approval, and those that may be waived using less rigorous management systems (see the additional discussions on limiting conditions for operation in Section 5.6.3 and control of impairments in Section 5.7.7)
- Have policies and practices in place for implementing and sustaining safeguards
- Understand risk tolerance and have an awareness of the residual risk associated with an activity (risk tolerance is addressed in Section 3.3.2)
- Perform periodic inspections and audits and have other means to measure and track compliance with significant standards
- Promote effective listening and learning that engage employees closest to the process in efforts to identify hazards and understand risk

5.4.2 Establish Standards That Support the Organization's Mission and Goals

Organizations with effective COO/OD systems establish appropriate standards and ensure compliance with standards through performance monitoring. Otherwise, any level of performance becomes acceptable, and the required level of performance seems arbitrary. Failure to set and enforce high standards for risk-significant activities dooms the COO system, leads to frustration, and generally sets the organization on the path to mediocrity. The COO system instills compliance with standards in every individual at a facility, and not just when a particular supervisor or manager is present. Organizations with effective COO/OD systems self-enforce the standards.

Most plants establish housekeeping standards; some enforce them better than others. At one plant, failure to properly dispose of oily rags used for a maintenance activity allowed the linseed oil to run onto a dryer, resulting in a fire that spread to combustible material that had spilled on the dryer and not been cleaned up. The fire continued to spread, resulting in significant financial loss.

Standards allow an organization to continuously evaluate individual and organizational performance against a fixed bar, and this provides early warning if performance appears to be slipping. That's not to say that standards must remain fixed; high-performing organizations periodically evaluate and adjust standards, in some cases eliminating ones that provide little benefit or adopting new ones. This process of revising or eliminating standards that provide little value presents special communication challenges. It should be very clear to stakeholders that this "pruning" process is not a compromise or lessening of the standard of care; rather it is part of an ongoing effort to establish fewer and more effective standards of performance that are widely understood. In addition, the bases for the standards should be clear to affected individuals.

Following are examples of standards found in the process industries:

- Processes are not intentionally operated outside of established operating limits, and if they are outside of established limits, prescribed actions are followed.
- Limiting conditions for operation (LCOs) are observed, and if an LCO is not met, the activity is not started or the process is promptly brought to a prescribed safe/stable state.
- If LCOs include minimum staffing, such a limit is observed under applicable conditions (e.g., startup).
- Decisions to defer maintenance, training, and other periodic activities are based on risk, not on budgets or resource availability.

These examples will not apply to all facilities, and by no means do they constitute an exhaustive list. They are included simply to illustrate situations to which standards might apply. Facilities should examine their formal and informal standards and thoughtfully determine which ones are guidelines, which ones apply under all but extraordinary conditions, and which ones must never be violated. All standards, their intended application, and, in particular, the "house rules" that must never be violated should be made crystal clear.

5.4.3 Understand What Can Be Directly Controlled and What Can Only Be Influenced

Individual requirements and goals should be based on what individuals can control or directly influence; avoid setting standards or holding people accountable for targets that they cannot control or significantly influence. For example, hold operators and maintenance personnel accountable for following specified procedures and methods, not for the results. Likewise, hold technical personnel accountable for establishing the procedures and methods necessary to achieve the desired results, and hold line management accountable for compliance with specified procedures and methods. Steven Covey includes this notion in his book entitled *The 7 Habits of Highly Effective People* (Ref. 5.12). Covey points out that we each have a "circle of concern" and a "circle of influence," with the first being larger than the second (i.e., we are often concerned about issues over which we have very little influence). According to Covey, focusing on issues/activities within our circle of influence in a proactive manner expands our circle of influence to include more of our circle of concern. Conversely, if we focus first on what we cannot influence, we become reactive, blaming others and convincing ourselves that things we actually could positively influence are completely out of our control.

A runaway reaction occurred when a new operator closely followed a written procedure and fully opened the valve in the line from the initiator tank to the reactor, precisely in the manner stated in the operating procedure. (More senior operators were aware of the need to crack the valve open to limit flow of initiator to the reactor until most of the reactant was consumed.) Although nobody was hurt, the runaway reaction did damage equipment, resulting in significant downtime and repair costs. Line management initially blamed the new operator, who had precisely followed the operating procedure as it was written, rather than asking why supervisors and managers, who could directly control operating procedures, had failed to maintain them current and accurate.

In this model, senior management at a facility is accountable for establishing and enhancing the organization's culture. Over the long term, organizational culture, rather than efforts by a handful of individual supervisors, superintendents, and managers, determines how the organization behaves.

Considerations for identifying accountabilities include the following:

- Determine whether goals will be met if the entire group adheres to procedures, training, practices, and standards. If not, or if an unrealistically low human error rate is required to meet goals, look for alternate, more reliable methods.
- Base accountabilities on what can be directly controlled. For example, production rates might be heavily influenced by the quality of raw materials supplied by an upstream unit. If the downstream unit is expected to consume both in-spec and out-of-spec product, it is unfair to hold only the downstream unit accountable for output goals. Rather, in this case, the accountability should be distributed between the two units.
- Review the level of effort and attention to detail required and compare them to the requirements for other work groups in the organization. For example, if operators are expected to self-check and peer-check to achieve very reliable and predictable results, the same standards should apply to all departments where noncompliance could lead to adverse consequences of similar severity.

5.4.4 Provide the Resources and Time Necessary to Complete Tasks Within Standards

Failure to provide the resources and time necessary to adhere to policies requires that personnel choose between adhering to policy and achieving expected results. Unless workers perceive that a policy violation creates grave danger, workers often tend to "bend the rules" a bit so that they can complete their assigned task.

> On February 9, 2001, the USS *Greenville*, a Los Angeles-class fast-attack submarine, struck the Japanese fishing trawler *Ehime Maru*. Shortly before the incident, the submarine's commanding officer (CO) ordered the officer on deck (OOD) to come to periscope depth (PD) within 5 minutes. The CO later told investigators that "He [the OOD] was so . . . slow, I knew that he couldn't get to PD in 5 minutes. It was my objective to give him a goal to work towards . . . I doubt that any of my experienced officers of the deck could have gotten to PD in 5 minutes." To comply with this order, the OOD skipped several important steps, including (1) performing certain maneuvers to determine whether there are surface ships nearby and (2) doing a periscope briefing. Ironically, these steps were skipped with the sub's captain and the Chief of Staff for the Pacific submarine command standing beside the OOD. Although the sub did not strike the *Ehime Maru* when it went to PD, this was part of a sequence of procedural errors that ultimately led to the death of nine people aboard the *Ehime Maru*. (Ref. 5.13)

It is frustrating to know what needs to be done and not be able to do it properly. Whether the cause is insufficient time, the unavailability of tools, or lack of manpower, the result is not only frustration but also potentially unsafe practices and incidents.

Processes involving hazardous chemicals require sufficient staffing and resources to ensure safe operation. When evaluating this attribute, consider the following:

- Are there times when a supervisor or leader prefers not to know how a task was done, essentially endorsing shortcuts or unsafe practices?
- When workers are observed taking a shortcut or misusing a tool in order to accomplish an assigned task, is the trend to look the other way (possibly because completing their work is important to getting the unit restarted or preventing a shutdown), or is the tendency to stop the work and incur the delay or cost necessary to perform the task in a safe/appropriate manner?
- During incident investigations, are employees unfairly held accountable for bending the rules even though it is widely known that the prescribed method for completing a task is impractical?
- Is staffing adequate and do staffing plans address modes of operation that require additional staffing, such as startup and other nonroutine operations?

- If a task cannot be performed properly due to poorly designed or worn-out equipment, is this brought to the attention of those who can remedy the situation?

5.4.5 Ensure Competency Across the Organization

COO tends to focus on individual performance and competence. However, individual competence cannot be sustained over time without organizational competence. An organization must continuously seek new information about the processes it operates and the tasks individuals perform; otherwise, it is doomed to repeat its mistakes and the well-documented mistakes of others.

The *Report of the Longford Royal Commission* documents the investigation of the September 1998 vapor cloud explosion in Longford, Australia. As one contributing factor, it identifies the transfer of engineers from the Longford site to Melbourne, resulting in an increased reliance on the competency of supervisors and operators. The individuals remaining in Longford were not fully trained on the hazards of low temperature brittle fracture; the resulting gap in organizational competency made the incident more likely. The explosion killed two workers, injured eight others, and interrupted the supply of natural gas to virtually the entire state of Victoria for two weeks.

In his book entitled *Lessons from Longford* (Ref. 5.14), Andrew Hopkins writes:

> *Historically a safety section at head office managed these low-frequency, high-consequence risks. These staff would oversee a number of facilities and ensure that the lessons learned about rare events were passed around. Esso should have known information on the rare catastrophic brittle failure of pressure vessels, as it was available from their parent company, Exxon. In 1974 and again in 1983, researchers from Exxon Research and Engineering Company published warning articles on these failures. As a direct result, Exxon had inserted into their hazard identification guidelines the requirements that special attention be paid to the possibility of brittle fracture.*

According to Hopkins, neither Esso's most senior manager in charge of risk assessment nor Esso's general manager was aware of the two articles that warned against brittle fracture hazards. Hopkins opines that "downsizing" of central safety staff and the decentralization of safety may have gone too far, and that failure to maintain the proper balance between central oversight and local control contributed to this incident.

The competency aspect of the Longford story applies to many other major process safety incidents. As discussed in Section 5.5.11, inadequate knowledge of piping design among workers assigned to fabricate a temporary jumper line resulted in piping failure and a vapor cloud explosion at Nypro's plant in Flixborough, England, in 1974, killing twenty-eight people (Ref. 5.9). In 1999, attempts to increase the concentration of hydroxylamine to unsafe levels at a facility in Lehigh Township, Pennsylvania, led to a chemical explosion that killed five workers. The severe instability of hydroxylamine at high concentration was well documented but not understood by those in charge of the facility at the time of the explosion (Ref. 5.15).

Process safety incidents often involve a lack of hazard awareness, failure to communicate information to key personnel, or organizational changes that cause the entire organization to forget what it previously knew. High-performing organizations continuously learn, freely and regularly share information with key personnel, and remember the lessons they have learned. Knowledge is treated as a valuable corporate asset. In his book entitled *Learning in Action* (Ref. 5.16), David Garvin identifies five "learning disabilities" that are often encountered in organizations. Any indications of these five learning disabilities should sound an alarm:

1. Blind spots – narrow focus, poor assumptions, disruptive technologies
2. Filtering – ignoring or downplaying information that does not fit into the existing paradigm
3. Lack of information sharing – ineffective sharing, information hoarding
4. Flawed interpretation – poor logic, emotional bias, hindsight
5. Inaction – inability or unwillingness to act

Highly competent organizations:

- Have designated technology guardians who are equally responsible for (1) maintaining and advancing knowledge in their subject area, (2) sharing this information with all people who may need it (completely rejecting the "knowledge is power" paradigm), and (3) documenting the information in a manner that it will survive and continue to grow long after the technology guardian moves on to other responsibilities
- Employ methods for sharing effective practices and tools between sites, such as a wiki (i.e., a repository of information that is developed and maintained by a user community) that includes items such as shift turnover logs and checklists that have been judged to be effective and compliant
- Consider the impact that staffing and/or organizational changes might have on competency, and formal action plans to address any issues that are identified
- Have a means to use company history to indoctrinate new personnel on why things are done in a particular manner, making training more pertinent, memorable, and "real life"
- Engage in frequent discussions that reinforce lessons learned from previous incidents and remind personnel of significant process hazards that are relevant to the job duties of each work group
- Develop and encourage an open work environment where comments and suggestions are welcome and acted on, regardless of who made them

5.4.6 Perform Critiques and Take Corrective Action

Any activity, whether it is flying a spacecraft or knitting a scarf, requires periodic evaluation and correction. Reliable plant operation is no different; evaluation, feedback, and corrective action are critical at several levels. Critique, as it applies to verification of individual training, is addressed in Section 5.5.4. Personnel effectiveness critiques (e.g., annual performance reviews) are outside the scope of this book. (Periodic performance counseling/feedback indirectly relates to COO/OD, but it is more closely related to human resource management systems.) The two areas that are explored in this chapter are (1) critique related to performance of infrequent but critical activities such as emergency response drills and (2) critique of the performance of management systems beyond the traditional audit.

> On September 23, 1999, the Mars Climate Orbiter crashed as it attempted to enter orbit. After loss of the spacecraft, investigators discovered that the trajectory errors were introduced by a software module that had been coded in the wrong measurement units. In addition to failing to identify the coding error during design and testing, NASA failed to detect signs indicating that there may have been a navigation issue with the orbiter. During its nine-month journey, propulsion maneuvers were required ten times more often than expected by the navigation team. However, since the course correction maneuvers continued to be successful, there was no formal inquiry to determine whether there was an underlying issue. (Ref. 5.17)

While one might argue that the crash of the Mars Climate Orbiter was more closely related to failure to review calculations and software, it clearly demonstrates how catastrophic failures can be preceded by subtle warning signs. A process safety event is often preceded by seemingly unrelated near-miss incidents. Since major loss events typically occur relatively infrequently, it is important to learn as much as possible from these so-called "weak signals." (And it is worth pointing out that what is a "weak signal" to one person might be a "red flag" to someone with a different perspective. Lack of a recent loss event cannot be the basis for discontinuing monitoring activities such as metrics collection, management review, and audits.) Likewise, since major emergencies are quite rare, emergency response drills are conducted for the explicit purpose of learning, and it is important to maximize their learning value by performing a critique and addressing what failed to go as planned.

Another form of periodic critique that was first introduced as an element of a PSM system in the CCPS's *Guidelines for Risk Based Process Safety* (Ref. 5.2) is the management review process. As described in Section 3.3.5, management reviews are scheduled periodically with the intent of conducting an honest self-examination of the performance PSM systems, even when there are no indications that anything is wrong. (Management review is described in more detail in Section 7.4.4.2.)

Efforts to identify and evaluate "weak signals" that often precede loss events and the performance of periodic management reviews have much in common. They both seek to learn from the past so that real corrections can be made. They both examine the effectiveness of recent mid-course corrections. They also seek to identify and evaluate improvement opportunities and to select the most appropriate opportunities based on requirements, risk significance, and good management practices.

5.5 PEOPLE

COO strives to promote high levels of human reliability. Despite technological advances in systems for communication, control, and error detection, errors continue to occur, resulting in greater risk. In fact, one might argue that the rate of change in these systems can exceed the rate at which humans can understand and use them, introducing new hazards, human failure modes, and potentially higher risk.

As described in Section 4.3, human errors are differences between acceptable and actual behavior or performance. They can result from (1) workers receiving incomplete, inaccurate, or conflicting written or oral instructions (or muddled communications), (2) failure to provide adequate training or work environments, and/or (3) structural breakdowns such as failure to detect and address worker fatigue or fitness for duty. The attributes addressed in this section include:

1. Clear authority/accountability
2. Communications
3. Logs and records
4. Training, skill maintenance, and individual competence
5. Compliance with policies and procedures
6. Safe and productive work environments
7. Aids to operation – the visible plant
8. Intolerance of deviations
9. Task verification
10. Supervision/support
11. Assigning qualified workers
12. Access control
13. Routines
14. Worker fatigue/fitness for duty

5.5.1 Clear Authority/Accountability

In the highly competitive global economy, companies often reduce headcount to achieve a target fixed cost. Although this is not completely negative and is often required for an organization's survival, it does increase the scope of responsibilities for personnel who remain and can create unclear roles and responsibilities. Some organizational changes are poorly managed, and some important responsibilities are not reallocated (or they are allocated to people with limited competence or inadequate experience), allowing critical PSM activities to languish. In other cases, supervisory positions have been eliminated, leaving members of self-managed work groups unclear about their authority, accountabilities, or responsibilities. Even in the absence of organizational change, conflicting instructions will lead to frustration and maybe to incidents. Anyone who has worked in industry for any length of time has heard a frustrated operator or mechanic turn to his supervisor, the superintendent, and the process engineer supporting the area and say, "Will you three just make a decision?"

According to the U.K.'s Civil Aviation Authority Safety Regulation Group, 70% of commercial airline incidents are due to human error. Traditionally, efforts to reduce human error emphasized the technical aspects of flying. However, investigations have found that training is not the only determinant of the outcome given an in-flight incident. Crew resource management (CRM), which focuses on effective decision-making, communication, and leadership, has been demonstrated to be a key factor in determining the ultimate result of a human error or equipment failure. Thus, considerably more emphasis is now placed on CRM during air crew training. (Ref. 5.18)

There should be a clear line of authority in any organization. Particularly under stressful conditions such as a process upset or incident, the operating staff should take direction from a single person – a designated person in charge – who may not be the highest ranking official present at the time. It is well established that the entire crew on an airplane answers to the pilot-in-command, rather than the highest ranking officer on board; likewise, instructions to operators should not be coming simultaneously from the shift supervisor, unit superintendent, and senior process engineer. This is not to say that technical personnel cannot provide advice and support to workers who do not work for them, but staff personnel should work through the line of authority, rather than directly issuing instructions. In addition, workers should be encouraged to question special instructions, and ultimately they should defer to the established line of authority if they are concerned about safety or some other aspect of an instruction.

In some cases, authority is handed off from one individual or group to another based on the mode of operation, work to be performed, or other events. For example, during an emergency, the operations group typically turns over control of a facility to the incident commander. When new facilities are built, there should be a formal handoff from the project team to the commissioning team, and ultimately to the operations group. Many facilities find it necessary to establish a policy regarding control of assets as they are shut down, deinventoried, prepared for maintenance, repaired, and subsequently returned to service.

Assigning accountability for tasks is also very important. Any task that is assigned to "all of us" and can be done by "any of us" is most often done by "none of us." It simply never gets done. Thus, in addition to lines of authority for day-to-day activities, clear accountability must be established for action items and important routine tasks.

Finally, accountability flows upward as well as downward. Managers who simply dole out assignments with no follow-up are often surprised when the tasks are done poorly, done late, or not done at all. Effective leaders go beyond simply looking at metrics related to how many tasks are not completed by the due date. They proactively track the progress of important activities, invest their time in face-to-face meetings to review progress and discuss concerns, and help remove roadblocks. More importantly, they feel some personal responsibility for completion of all tasks assigned to their organization: "If anyone fails, we all fail." In short, leaders feel accountable for everything the organization does or fails to do.

Organizations that have established clear lines of authority and accountability ensure that:

- Accountability is assigned for work activities and projects/tasks
- Workers understand the lines of authority at all times
- Technical and other support personnel understand the lines of authority, the extent of their authority is clearly defined, and they work within the established system
- Accountability flows upward as well as downward; managers make assignments and follow up to ensure that the tasks are completed

5.5.2 Communications

Complex operations require that activities be coordinated among individuals within work groups and among work groups. This is particularly important for facilities that process hazardous materials because any activity that involves the coordinated activities of multiple people depends on effective communication.

Communication involves some or all of the following elements: (1) a sender, (2) a receiver, (3) a message, (4) a medium, and (5) feedback and confirmation methods. COO focuses on ways to minimize errors involving each of the five elements. Examples of ways to reduce communication errors include:

- Structuring messages in a standard format, which alerts the receiver if important information has been omitted or if an entire section is skipped
- Providing written instructions or procedures so that they can be reviewed by the sender, receiver, and others to reduce the likelihood that part of the message is missing, confusing, or wrong
- Establishing protocols to repeat key parts of the message back to the sender, which is particularly important for verbal communications
- Using structured protocols, checklists, and logs to supplement written instructions
- Preceding special or nonroutine activities with a pre-job briefing for the entire work crew and a field walkdown by people with appropriate knowledge/experience

Most paper mills use chlorine dioxide (ClO_2) to bleach pulp. The R10® ClO_2 process requires the mill to periodically take a short outage to "boil out" the reboiler on the ClO_2 generator. In addition to improving the efficiency of the generator, this provides an opportunity to perform maintenance activities on other parts of the ClO_2 process. After the boilout is complete, liquor is transferred back to the generator (reactor) and the process is restarted. As the reactor and other parts of the process approach steady state, the salt cake filter is brought online and a slip stream of liquor is routed to this equipment to maximize chlorate yield.

During a routine boilout, mill management elected to do some maintenance work on the salt cake filter system, requiring that the system be isolated, drained, rinsed, and opened up. This work was completed without incident. However, day-shift personnel had failed to close one of the drain valves on the line to the salt cake filter. When evening-shift personnel arrived at 6:00 p.m., day-shift personnel reported that they had restarted the generator and that it was almost time to bring the salt cake filter back online. During the turnover, there was no mention of the work done on the salt cake filter, and there was no record of it in the shift log. Assuming that all valves were in their normal position, night-shift personnel brought the salt cake filter online in the standard manner, causing an immediate release of ClO_2 solution. Fortunately, the release was promptly detected and the line was isolated with no injuries or harm to the environment.

Although this incident resulted from several management system gaps (e.g., failure to properly "undo" a double block and bleed arrangement, no checklist to ensure that all valves are properly aligned prior to restart), it is highly likely that if night-shift personnel had known of the maintenance work, they would have verified the system's status before placing the salt cake filter in service.

5.5.2.1 Checklists and Logs

Failure to communicate during a transition can leave the oncoming shift team with incomplete information and vulnerable to mishaps. For example, if there is no specific list of topics for a shift handover, the oncoming operator may clearly understand and remember every word spoken by the departing operator, but the oncoming operator may not receive other important information simply because the departing operator was tired and forgot to mention it. Tools that help promote communication during transitions include checklists and logs. Due to the continuous nature of most facilities in the process industries, there is a face-to-face meeting at shift change as the oncoming operator reports to the control room or other work area. However, an effective shift turnover will not happen on its own; a management system is needed to guide the discussion. Checklists, coupled with a log of activities, help ensure that important information is passed along in written form to remind the oncoming operator of the initial status of the unit.

Methods to improve communications within an operating unit include the following:

- Using forms to provide structure, particularly for shift logs
- Establishing protocols for verbal communications to repeat key parts of the message back to the sender. This becomes even more important in high-noise areas or when using two-way radios or other methods where sound may be garbled.
- Providing a current display of all equipment that is out of service
- Having outgoing personnel verbally review the information in the shift log with relief personnel
- In shift logs or similar notes, clearly designating:
  - routine operations
  - transition operations (e.g., changes to produce a different product or switch to a different raw material source)
  - nonroutine operations (e.g., maintenance activities or temporary injection of an additive to reduce fouling)

While checklists tell what was (or was not) done, they often lack details about when and how activities were performed. Well-designed shift logs that provide structure and use consistent terminology are a much better means to capture this information. Shift logs are used to record important routine activities and all nonroutine activities/situations. Operator round sheets typically document the status/condition of field equipment every few hours. Together, checklists, shift logs, and round sheets form a solid foundation for routine shift-to-shift communication. Facilities should establish a policy regarding when each should be used and what information should be recorded. In addition to helping the oncoming shift understand the status of the unit, this information is often a critical part of the historical data needed for process troubleshooting and incident investigations.

Checklists can also reduce errors during low-frequency but very important work activities. For example, facilities often build checklists for items that need to be verified following a turnaround. In most cases, the turnaround group exercises significant control over activities that occur while the unit is down and is responsible for turning the unit over to operations ready for restart. However, the operations group is typically responsible for checking valve lineup. Although this example is relatively straightforward, most turnarounds involve a large number of complex, interrelated tasks, and there is significant opportunity for error if handoffs are not formalized. Properly developed checklists that are reviewed, updated, and supplemented with special checklists for special projects help ensure that nothing falls though the cracks during the ensuing startup.

In a similar manner, checklists document the distribution of responsibilities for new units between the construction, commissioning, and operating teams. These checklists can be reviewed well in advance to make sure that they include all critical activities. Activities can be assigned to various groups, and each group can use the checklist(s) to maintain the status of assigned items.

One common pitfall of checklists is that they can become so familiar or routine that users fail to read each item; they simply check (or initial) each block after the task is complete. This is an example of an OD issue; compliance with standards is addressed in Section 6.2.3. Another pitfall is that by design, they tend to focus the user. Although this is the intent and it is normally helpful, it can cause the user to overlook a significant hazard or concern due to myopic focus on each checklist item.

5.5.2.2 Required Reading

Another form of transition is when changes are made to process conditions, procedures, or the physical plant. In many cases, workers can quickly adapt to changes as long as they know what has been done. To help facilitate this communication, many facilities establish a required reading file (i.e., a group of documents, such as change approval forms, that must be read by everyone in a work group at or near the start of each shift). This can also take the form of a note in the operator's logbook, an e-mail, a bulletin board posting, or similar one-way communication. Effectiveness of these systems is normally directly related to personal accountability. E-mails with no acknowledgement or postings on the bulletin board typically have low levels of effectiveness. Required reading with a block for each person to initial that he or she read and understood the message is usually more effective. The use of any of these methods within the COO system, supplemented by routine questioning from supervisors, not only causes personnel to review the material, it also emphasizes the importance of the communication process. This reinforces personal accountability, which is an important aspect of OD.

5.5.2.3 Special Written Instructions

Oral instructions, especially when they concern nonroutine tasks, should be supplemented by written instructions. Most facilities require that any temporary or special instructions be written to help ensure that nonroutine tasks are performed safely and reliably. As one common name implies, "night orders" are written communication between day-shift and night-shift personnel who may not be in the same location when the instructions are issued and/or executed. These orders should be based on, and reinforce, standing policies and procedures. In addition, night orders should have limited duration. A management system should trigger routine review of the current night orders, forcing their cancellation or incorporation into standard procedures or policies.

5.5.2.4 Oral Briefings

It is important to ensure that all personnel involved in the work are aware of the overall plan. A pre-job briefing provides the opportunity to gather the entire team, review the task and the associated hazards, emphasize critical steps, and address concerns expressed by any of the team members. Ideally, each person involved would have carefully studied the instructions, be well aware of the hazards, and understand the critical steps prior to the activity; however, this is often not the case. A face-to-face meeting allows the person in charge to review the steps with the involved personnel and, more importantly, to examine their body language, evaluate their questions, and measure the level of engagement from each team member, which provides a good indication of each individual's level of comprehension. Likewise, face-to-face discussions when work is complete help ensure the correct post-task equipment status and help identify opportunities to improve the plans and procedures. This feedback is an important part of efforts to continuously learn and improve.

5.5.2.5 Summary

Effective communication is vital to any organization or activity that involves more than one person. The message, media, and rules to ensure that the message was properly sent and received should be based on how important it is for the message to be received properly and the recipient to take the proper action. Communicating in multiple formats and/or at multiple times will improve reliability. For example, providing a written instruction or checklist for a special one-time task is normally preferred, but it may limit feedback. Conversely, issuing oral instructions allows for immediate feedback and verification of understanding, but the receiver may subsequently forget some of the message. Holding a pre-job briefing to review written instructions or a checklist with the crew that is about to do the work significantly enhances performance compared to using either communication method by itself.

Tools for effective communication include:

- Standard checklists, which help ensure that critical information or steps are not omitted
- Shift logs, which provide a chronological record of activities and help highlight any nonroutine activities or observations
- Specialized checklists, which help allocate responsibilities among work groups and ensure that all tasks are performed prior to key events, such as startup following a turnaround
- Required reading, particularly for communicating details of minor changes or informing potentially affected personnel of incidents, results of incident investigations, or results of hazard analyses
- Written instructions, particularly for performing nonroutine tasks that are not otherwise addressed in established procedures or policies
- Pre-job briefings
- Post-task debriefs

5.5.3 Logs and Records

Although logs and records are a form of communication, they merit separate discussion. They are important, even when the person making the entry has no immediate reason to believe the information is significant. Systems and equipment often provide plenty of warning signs prior to catastrophic failure, but the early signs may be observed by different people. Evaluated in isolation, each sign may be dismissed as insignificant. Records and logs, whether paper or electronic, allow workers to compare existing operating parameters to historical data to evaluate changes and trends and to determine the rate of change.

> "C" shift personnel noticed that the control system was slow to respond, but they failed to communicate the problems to "D" shift personnel. Twelve hours later, "A" shift personnel returned after their seven-day break and noted the same response-time issue with the system. However, nobody logged in or reported the problems because they seemed minor and sporadic. Several hours later, shortly after "B" shift personnel arrived, a catastrophic control system failure occurred.

Logs/records should be kept for (1) process conditions, (2) major equipment items, (3) unusual activities, and (4) incidents or unplanned events. Many processes are controlled by digital electronic devices with the ability to record large volumes of data. These data become invaluable when troubleshooting and should be frequently monitored by operators to spot slowly developing trends. It is also good practice to have field operators record a set of readings every few hours using field gauges and to compare them to board indications. Logs or round sheets should include upper and lower limits to facilitate comparison to unsafe/unacceptable limits. Otherwise, field operators might notice that a reading is abnormally high or low, but fail to recognize that it is outside of a limit. Ideally, these readings should be trended over time to enhance the ability of personnel to spot patterns of change in a particular parameter, possibly indicating calibration drift in a sensor that is associated with a control function. Field readings can also be used to identify rapid changes in process conditions that were not identified by the control system, possibly due to a failure of the control system. In addition, some critical information comes from direct observations of conditions in the field; sounds and smells can be very important and may provide information that the numbers alone cannot. (See Section 5.7.2 for more information on equipment monitoring.)

Most facilities do a good job of reporting and investigating major incidents. The COO system should also require diligent reporting and investigation of near-miss events – ones that could have had serious consequences if the safety systems had failed to work. However, the best COO systems require a record of the unexpected events, even when no loss event occurs. For example, a testing failure for a safety instrumented system is often viewed as a near miss, even though the stated objective of the testing is to find and fix defects. Establishing a means to easily report these types of events in an open manner and tracking the results can help identify chronic failures or failures that indicate weaknesses in management systems, process design, and process operation.

5.5.4 Training, Skill Maintenance, and Individual Competence

Since human error is a significant contributor to many major incidents, initial training and refresher training are critical elements for high-performing organizations. In systems where a high degree of hardware redundancy minimizes the frequency or consequence of single component failures, human error can account for over 90% of system failures (Ref. 5.19). This section examines training – actions taken to teach workers how to perform new tasks and procedures or to refresh their knowledge of them. Key characteristics of an effective training program include (1) content that is based on an assessment of trainee needs, (2) sessions that are provided at the right time (just-in-time or periodic), and (3) verification of trainee performance or understanding that is based on an established standard.

On January 30, 2007, a junior technician was assigned to transfer liquefied propane from an existing storage tank to a newly installed replacement tank at the Little General Store in Ghent, West Virginia. When the technician removed a plug on the liquid withdrawal line, liquid propane was released. The line was rarely used, and there was a safety feature – a telltale hole drilled through the threads – that should have warned the technician that the valve under the plug was leaking through. The U.S. Chemical Safety and Hazard Investigation Board (CSB) determined that the technician was likely unaware of this safety feature due to inexperience and lack of training. Ignoring the telltale leak and removing the plug led to a large release, eventually resulting in an explosion that killed four people and injured six others. (Ref. 5.20)

Refresher training is particularly critical when (1) the worker is asked to perform a task with little opportunity to review the procedures, (2) a high level of performance is critical, and (3) the task is not performed very often (or the task is being performed for the first time). Even if only one or two of these conditions apply, the task is a good candidate for inclusion in refresher training. In many cases, it makes sense to establish an event-driven training interval that aligns with when the activity occurs. For example, refresher training on shutting down for turnaround and starting up after turnaround should be scheduled shortly before the turnaround, versus a calendar-based interval that does not coincide with the turnaround schedule.

Verification of understanding can take three forms: written test, verbal test (normally via structured interview), or performance demonstration. Each method has its strengths and weaknesses. Written multiple-choice tests can become more a measure of the trainee's ability to take a test than an evaluation of knowledge. Unstructured verbal tests can be very unreliable, with the depth of questioning varying widely from one trainee to the next.

Performance demonstration is typically the best means to evaluate understanding of skill-based training. It normally consists of a series of experiential learning events: "watch me" followed by "I watch/coach you" followed by "I watch/test you." While this approach helps ensure that the trainee is capable of doing the task, it is also a very critical step in reinforcing the training. Properly designed tests (1) motivate the trainee to pay closer attention and practice new skills and (2) provide an opportunity to coach trainees. Regardless of the evaluation method used, the tasks, the conditions under which they will be performed, and the minimum performance standards should be documented and well understood. This helps ensure that standards are valid and promotes a sense of fairness in the performance evaluation process.

Often, training programs are laid out as a series of modules, with no event that ties all of the pieces together. Just as engineering school curricula often culminate with some sort of "senior project" that requires students to demonstrate their ability to apply knowledge from many courses to a single problem, a final qualification process at the end of a training program can help determine whether trainees fully understand and are able to apply the training they received. This process is typically a structured interview with a knowledgeable manager or with a group that includes peers and one or more supervisors/managers. Sometimes it involves a structured interview conducted in the field by the unit superintendent, with the intent of evaluating both the competence of the trainee and the quality of the training program.

Finally, training programs should be tailored to the objectives. This might include teaching the trainee to (1) follow a procedure (rule-based performance), (2) adeptly complete an activity (skill-based performance), or (3) diagnose a situation and take appropriate action (knowledge-based performance). Within established limits, there is a place for each approach. In addition, training should reinforce process limits and LCOs; and, regardless of one's knowledge, these limits should not be intentionally breached.

Consider the task of learning to drive a car. The student driver must actively process each input, decide on a course of action, and then respond. For example, as a right turn approaches, a new driver may go through a mental checklist of the rules (check for traffic, activate turning signal, move to right, slow down, etc.). As experience and competency improve, a right turn (and driving in general) requires less conscious thought (skill-based). In a much more complex environment like a chemical plant, some activities (such as following a checklist for a batch procedure) should remain rule-based. For frequently performed tasks (such as starting a centrifugal pump), training should seek to achieve skill-based behavior for trainees. Training for other activities (such as troubleshooting) should provide trainees with the knowledge, skills, and abilities required to perform knowledge-based tasks (see Chapter 4).

Effective training programs have the following features:

- Training is as realistic as possible, using simulators where they exist.
- Initial training is provided just-in-time.
- Refresher training is provided periodically and in a timely manner.
- Trainees are evaluated against a specific standard that closely simulates real-world conditions.
- Trainees receive coaching and feedback that help them improve performance.
- Training promotes rule-, skill-, or knowledge-based responses, depending on the anticipated nature of the work.

Effective training programs do not stop with teaching "what to do"; they move on to explain how the process works, why the intended action works, what might cause it to fail, and what may occur if the action is not performed properly. This understanding helps personnel quickly recognize and react to process upsets or abnormal conditions. A thorough understanding of the process is a precondition for "thoughtful compliance," which is addressed in the next section.

5.5.5 Compliance with Policies and Procedures

One of the cornerstones of COO/OD is compliance with policies, procedures, and established practices. Any shortcuts, particularly shortcuts sanctioned (or intentionally ignored) by management, can undermine the entire program. This, in turn, requires that procedures be maintained current and accurate and that training match the procedures.

> Plant procedures required that workers use supplied air masks and acid suits when testing for leaks after attaching an unloading hose to a chlorine railcar. However, the crew normally assigned to perform this task had become very adept at making the fittings, and rarely encountered a leak, particularly one of any significance. At some point, they quit wearing the hot, uncomfortable safety gear, until one day there was a large leak when they opened the valve on the car. Although the crew escaped without injury, the resulting chlorine cloud prevented them from quickly closing the valve on the railcar, leading to a large release.

Compliance with procedures helps ensure quality, overall reliability, and organizational effectiveness. For example, crews on commercial airlines rarely work together for more than a few days during any given year. However, the airlines have an enviable record of safety and cabin-crew efficiency, due in part to the fact that procedures have been developed for each task, from preflight checks to aircraft cleaning and servicing, and there is a well-established cultural mandate to follow procedures.

Finally, there is an expectation of thoughtful compliance. That is, the person assigned to perform the procedure, as well as his or her coworkers and supervisors, are expected to be on the lookout for hazards that are not fully addressed in the procedure. Unusual or unexpected feedback or conditions, or any similar signs that the procedure is not adequate or appropriate for the conditions, should be evaluated closely. Under these conditions, a decision to continue simply because the procedure does not explicitly say "stop" is clearly not thoughtful compliance. Rather, the operation should be promptly brought to a safe state until the issue can be evaluated and resolved.

Organizations that promote compliance with procedures exhibit the following traits:

- Personnel who interrupt work due to an anomaly are congratulated on their attention to detail rather than chastised, regardless of whether their concern is borne out.

- When improvements are suggested to procedures, policies, or practices, a formal MOC system is used to evaluate the suggestions, and the scope of the review includes both normal/anticipated conditions and credible abnormal conditions.
- Management and supervision diligently ensure that improved outcomes are not based on shortcuts or unauthorized modifications to established procedures, policies, and practices, and they take appropriate steps to discourage unauthorized changes, regardless of the degree of ingenuity or outcome.

5.5.6 Safe and Productive Work Environments

People often form their first impression of a facility based on housekeeping. A clean, organized facility generally "feels" more safe and efficient. In most cases, this is a very correct assumption, but due more to correlation than causation. In other words, the same cultural factors that influence workers to maintain a clean and orderly work area normally influence them to adhere to other standards as well.

The CSB published an investigation report (Ref. 5.21) concerning three separate dust cloud explosions that claimed a total of fourteen lives. According to the report:

> *On January 29, 2003, a massive dust explosion at the West Pharmaceutical Services facility in Kinston, North Carolina, killed six workers and destroyed the facility . . .*
>
> *West produced rubber syringe plungers and other pharmaceutical devices at the facility. In the rubber compounding process, freshly milled rubber strips were dipped into a slurry of polyethylene, water, and surfactant to cool the rubber and provide an anti-tack coating. As the rubber dried, fine polyethylene powder drifted on air currents to the space above a suspended ceiling.*
>
> *Polyethylene powder accumulated on surfaces above the suspended ceiling, providing fuel for a devastating secondary explosion. While the visible production areas were kept extremely clean, few employees were aware of the dust accumulation hidden above the suspended ceiling . . . [nor] were aware of the . . . hazards of combustible dust.*

This incident is one example where maintaining what appeared to be a clean work environment was not the same as maintaining a safe work environment. Although there are many reasons to maintain a clean work environment, this is not always a sufficient condition for safe operation.

While good housekeeping may be a worthwhile goal and may be critical to worker or product safety in certain environments, it cannot be an end unto itself. In the process industries, cleanliness and lack of clutter do not ensure safety; however, clean and orderly facilities are certainly safer than facilities where housekeeping standards are not enforced.

Effective organizations understand the factors that lead to a safe and productive work environment, and they take steps to ensure that those factors are in place. For example, dust can create hazards ranging from slips and falls to fatal lung diseases to multifatality explosions. However, a safe and productive work environment requires more than just a visibly tidy facility.

Maintaining an orderly work environment also helps improve productivity. In addition to creating safety hazards, clutter and disorder cause people to waste more time looking for things, take shortcuts, or use the wrong tool or part. Many companies have adopted a "5S" program as part of their total quality management effort. This approach fits into a COO program, and it includes the following five steps:

1. Sorting: Keep what you need and discard the rest.
2. Set in order: Arrange tools, materials, parts, and other items in a manner that makes them easy to locate and maximizes efficiency.
3. Sweep: Maintain a clean work environment.
4. Standardize: Standardize materials, procedures, and routines.
5. Sustain: Maintain and enforce the standards that make up the 5S program.

Workers tend to take more pride in their work in a clean, orderly environment, and pride in workmanship can positively influence other, more tangible aspects of the operation. Regardless, a cluttered, dirty, or poorly organized work environment is unsafe and unproductive.

Activities that help promote efforts to maintain a safe and productive work environment include the following:

- Housekeeping inspections that extend to all areas of the facility, including offices, laboratories, guard houses, maintenance shops, and other work environments
- Clearly marked and appropriately located waste containers to help prevent mishaps or environmental issues due to improper waste disposal
- Orderly stock rooms, where parts are properly segregated, appropriate cleanliness standards have been established and are enforced, and access is controlled
- Clear expectations for all workers, including office personnel, regarding what they must do to maintain a safe and productive work environment

5.5.7 Aids to Operation – the Visible Plant

Chapter 4 discussed several ways to improve human performance. Vision tends to dominate the human senses ("seeing is believing"), so making plant conditions visible is one of the most effective ways to improve human performance. The visible plant includes labeling, signage, and color-coding, but it goes far beyond that to provide visual indications of abnormal conditions.

The introduction of distributed control systems and programmable logic controllers ushered in many advances in process control, but these improvements have made some aspects of the plant less visible. Old analog panel boards depicted the physical arrangement of the entire process, and a quick scan by an experienced operator was all that was required to monitor the process and identify any unusual conditions. Alarm panels were normally located toward the top of the panel board, and one of the most noticeable things in the control room was a light on the alarm panel board. Since the number of conditions that could be alarmed was limited, these alarms usually indicated a significant process upset, and operators generally responded to these alarms in a timely manner.

In more automated facilities, a greater burden is placed on the designers of the control and alarm systems to prioritize information for personnel and provide prompts to lead them to the right data. Facility personnel should identify the information (data) that is needed under each plant condition, then determine how that information will be provided to personnel (displays, alarms, logs, verbal communications, observation, etc.) in a way that makes the plant visible. Finally, a holistic review should be performed to ensure that personnel can appropriately understand and prioritize the information.

Properly designed visual displays use colors and symbols to indicate which equipment is active, whether any interlocks are active, or whether other abnormal conditions exist. For batch or step-wise processes, the control system also indicates the process step in a standard location on the operating console.

A specialty chemical manufacturer operates its facilities inside buildings to protect against product contamination. Product purity requirements are very high, and in several cases, flammable solvents are used for fractional crystallization and chromatography separation processes. One highly flammable solvent is filtered as it enters the main production building, and the filter is located in a normally unoccupied room of an otherwise highly occupied building; therefore, a solvent sensor was installed near the filter to alert operators to any loss of containment. However, the solvent sensor tripped every time the filter was changed. To prevent nuisance alarms, the facility modified the filter change procedure to first cap the sensor, change the filter, wait a few minutes to allow the solvent vapors to clear, and then remove the cap from the sensor. In at least one case, the operators neglected to remove the cap from the sensor, and the gasket on the filter subsequently failed, resulting in a large release of solvent inside the building. Fortunately, the flammable vapor cloud did not find an ignition source.

After this incident, the facility modified all procedures involving capped sensors. The number of caps available in each process area is now tightly controlled, and the caps are stored in a highly visible place in the control room. If a cap is not in its location, a large red circle is visible on the wall, alerting operators to locate and remove the cap (unless they are certain that a filter change or similar activity is underway).

The visible plant quickly informs the operator of abnormal conditions, including:

- Safety systems that are bypassed, impaired, or out of service
- Process alarms that are disabled
- Process conditions that are not within established limits
- Backup systems that are unavailable
- Special maintenance or testing activities that are currently in progress

5.5.8 Intolerance of Deviations

> One well-documented example of a widely tolerated deviation was the shedding of foam from the external fuel tank that caused the space shuttle Columbia disaster. NASA's standard called for zero foam shedding, yet there had been more than sixty documented instances of foam shedding prior to the launch of Columbia in January 2003. As an organization, NASA had grown to accept this risk but had never revised its foam shedding standard. (Ref. 5.22)

Deviations can take many forms. As previously described in this chapter, people can deviate from established procedures, policies, and practices. Processes can be operated outside of established limits, either intentionally or unintentionally. Equipment can be operated with safety systems out of service. Each of these situations involves the loss of a layer of protection, with the potential for catastrophic consequences.

COO/OD emphasizes conducting activities in a prescribed manner every time, without deviation. Many of the systems described in this chapter and throughout this book promote this concept. Deviations, whether they are known to the entire organization or an individual employee, cannot be tolerated. This section focuses on deviations that are known to a wide number of people. Section 6.2.3 focuses on individuals following procedures and standards.

An effective COO system will help establish an organizational culture that refuses to tolerate deviations – at least not very many or for very long. When faced with a deviation, the organization's response is "that's not like us" or "that's not how we do things," and actions are taken to manage risk effectively and address problems in a timely manner.

A critical step in promoting intolerance of deviations is to make the standards very clear. For example, almost all PSM systems include a requirement to describe safety systems in operating procedures. This is partly so that operators know how to react when an important system trips; however, it also helps operators and others hold management accountable when these systems are out of service. Accountability for conforming to standards works both ways: management sets expectations for workers, but workers should also hold management accountable for correcting deviations.

The following conditions might exist on an offshore oil production platform:

- Two gas detectors in Zone 1 are inhibited due to false alarms.
- Inspectors have found critical wall thinning on a hydrocarbon line in another zone, so instructions have been issued for enhanced operator monitoring in that area.
- During the test of safety systems last week, the main fire pump did not start, but the backup did start and a firefighting boat is maintaining watch in the area.
- Two action items from process hazard analyses related to relocating blowdown lines to a safe location have not yet been addressed and are currently past due.
- Hot work is planned for Zone 2 today, which is adjacent to Zone 1.
- Tests of the safety shutdown valves for the platform have been delayed for three months.
- Due to demand, some of the flow lines on the platform have been operated above their erosional velocity for the past month; this was an authorized change, and the lines are being closely monitored.

The risk associated with each of these conditions might be considered tolerable. In fact, a platform might be operating today under similar conditions and still not be likely to experience a loss event. However, when these facts were input to the quantitative risk analysis of the facility, the model showed that the cumulative effect increased risk by more than a factor of 100.

Following most major process-safety incidents, people often comment on how unlucky the facility was that several seemingly unrelated conditions were simultaneously in place. Actually, too often there is a common cause: the facility tolerates deviations and fails to correct them in a timely manner.

Risk is a function of frequency, which often depends on exposure (the fraction of the time that the postulated scenario might occur). Deviations should be avoided if possible and corrected immediately once they do occur. Although some might argue that it is safe to operate an automobile with one headlight burned out, it would be foolish to suggest running the second headlight to failure. Promptly replacing a failed headlight reduces risk by reducing exposure to failure of the remaining headlight while driving. Likewise, taking prompt action to address deviations and promptly revert to standard operating procedures, reestablish standard operating conditions, and restore equipment to normal operating conditions reduces risk by reducing exposure to abnormal conditions.

Organizations that refuse to tolerate deviations exhibit the following traits:

- Standards are well understood throughout the organization.
- Management sets the example by promoting and enforcing standards.

- When standards are changed, the reasons are made clear and the rationale is sound; standards are not simply changed because they are inconvenient or adversely impact short-term financial results.
- When it is necessary to operate outside of standards, there is a process for authorizing the exception, the process appropriately considers risk, and it is strictly followed.
- Regardless of the expected outcome or whether the activity is directly related to their job responsibilities, members of the organization resist the temptation to look the other way when a procedure is not being followed.
- When standards are violated without authorization, individuals who should have been enforcing the standards, authorized the activity, and actively participated are all held accountable.
- Workers hold peers accountable for adhering to policies, procedures, and standards.

5.5.9 Task Verification

Well-designed systems are tolerant of human error. As described in Section 4.3, people are fallible, and even the best workers err. Therefore, highly reliable procedural controls are valuable in identifying and fixing errors before they initiate an incident.

> The final two steps in the production of a specialty intermediate chemical are a solvent wash followed by vacuum drying. Once the product is dry, it is stored for future use, sometimes remaining in the warehouse for several months. The primary reason for the wash is to remove an impurity that reacts slowly but exothermically with the bulk material, resulting in an accelerating temperature rise. If sufficient impurity is present, the material can reach its autoignition temperature.
>
> Due to safety and product quality concerns, the facility assigned a second person to verify that the operator completed each step, including the solvent wash step. However, over time this evolved into "checking the paperwork" several days after each lot was produced. Worse yet, the verifier tended to "fix the paperwork" rather than investigate anomalies.
>
> While investigating a warehouse fire that did not result in injuries but did result in a multimillion dollar property loss, it was determined that one lot of this material was stored in the vicinity of where the fire started. The incident investigation team confirmed that (1) the process data historian showed no flow in the solvent line to the wash tank at the time the wash was supposed to be conducted and (2) the verifier had written the operator's initials on the batch sheet, indicating that the wash had been performed based solely on the operator stating that he believed it was done.

Almost all permit systems have a verification step. For example, to obtain a hot work permit, the fitter, welder, and operator typically prepare the area for hot work and the supervisor inspects the worksite, using the permit to confirm that all requirements have been met. The supervisor then authorizes the hot work. Similar systems are commonly used for other maintenance and nonroutine work, typically because this type of work involves special hazards and presents potential communication issues between work groups.

If the consequences of a human error are unacceptable, consider adding an independent verification step. Self-checking is a useful practice, but someone other than the person who performed the work should verify critical steps. To help prevent complacency, both the "doer" and verifier should clearly understand why specific steps are critical. Basing the verification on a single question that effectively asks "Did you do everything right?" and signing off solely on the basis of the answers provided by the doer are risky. An even worse practice is reviewing the checklist and filling in blanks that were not completed without verifying that the task was completed correctly.

Another alternative is to use hardware-based verification. Often referred to as error-proofing, this method uses an engineering feature to enforce the proper behavior. For example, to help prevent electrocution, buckets on motor control centers are fitted with mechanical interlocks to prevent the door from being inadvertently opened when the breaker is energized.

Another form of verification is planned job observations. When new procedures are implemented or as part of periodic verification for existing procedures that are deemed critical to safety, supervisors or other assigned personnel directly observe each worker assigned to perform the procedure, correcting workers when they make errors and providing feedback on the error rate so that appropriate adjustments can be made.

In addition to formal verification processes, it is important to confirm that workers self-check. For example, if the operation is to make up a batch of additives containing two bags of catalyst, three pails of an additive, and a liter of dye solution, procedures should require that the operator start by moving all of the required materials to the manway for the batch tank – no more, no less. A checklist can help ensure that materials are added in the proper order. Often, data such as lot numbers are recorded on this sort of checklist, forcing the operator to fill it out as materials are added to the tank. Typically, the final step is to inventory the number of empty containers to ensure that nothing has been left out and that no extra materials were added to the tank. Likewise, physical error-proofing approaches should be considered for parts and tools. For example, limiting the maintenance materials in the area to only what will be needed for the work, and using carts with a designated location for each tool, can clearly indicate whether a tool was left inside the equipment or a part was not installed.

Checklists have proven to be powerful tools in a wide variety of situations. As described in Chapter 2, a recent study by the World Health Organization shows that using a simple nineteen-item checklist prior to surgery reduces the rate of death following surgery by almost 50% and the rate of complications by over 35% (Ref. 5.23).

Procedures should state the expected response from any activity, and operators should be trained to confirm the response. For example, after starting a pump, operators typically confirm that the discharge pressure is within the normal range and visually confirm that valves are properly aligned. If the expected response is not observed, the activity should be suspended (or the process or equipment reverted to a safe state) until the discrepancy can be resolved.

Ways to promote verification efforts include the following:

- Include the expected system response in procedures, and train operators to confirm that the system responds as expected after each action step.
- Use hazard analysis methods to identify high-risk operations or situations where a single human error can trigger a chain of events that can lead to an unacceptable outcome; consider implementing an effective verification step to help identify and correct mistakes in a timely manner.
- When human error presents a significant or special hazard, implement verification processes such as planned job observations.
- Require independent technical reviews and verification for all activities that are critical to process safety, such as chemical analyses, engineering calculations, design work, hazard and risk analyses, and incident investigations.
- Implement error-proofing and other visual means to verify that activities are properly executed.

5.5.10 Supervision/Support

Among many other duties, supervisors assume the roles of trainer, coach, leader, coordinator, supporter, technical authority, evaluator, and enforcer. In many respects, supervisors are the real-time risk managers for production facilities. Day by day, hour by hour, they make decisions, apply resources, direct activities, and enforce policies. A COO system depends on such strong leadership.

> While a hydrocracker unit was being restarted at a refinery in Grangemouth, Scotland, early on the morning of March 22, 1987, an automatic shutdown was caused by a spurious high temperature reading. After resolving the problem, personnel decided to hold the unit on gas recirculation and restart it once the more experienced day-shift supervisor arrived. The usual shift handover occurred at 6:00 a.m. and included an instruction that the unit remain on gas recirculation pending the arrival of the day-shift supervisor. For some reason, the operating crew proceeded with the startup at 7:00 a.m. and began by manually opening the liquid level control valve that isolated the high-pressure separator from the low-pressure separator to warm the lines and confirm that they were not blocked. The control room operator on duty was not trained in this method, he had not previously performed this task, and he failed to wait for the day-shift supervisor. Shortly after the high-pressure separator level control valve was manually opened, the remaining liquid was transferred to the low-pressure separator, followed by high-pressure gas. This burst the low-pressure separator, leading to a large release of hydrocarbons and a vapor cloud explosion that killed one worker and scattered pieces of the low-pressure separator (weighing up to three tons) over a distance of a half-mile. (Ref. 5.6)

In addition to maintaining the corporate memory, leaders establish the culture and conscience of an organization. Effective leaders embrace the notion that they are responsible for everything that their group (or organization) does and fails to do.

Chapter 3 described the need for leadership commitment. An effective COO system depends on effective leadership at all levels of the organization. Although some people seem to have an inherent ability to lead, leadership is a trait that can be learned, practiced, and honed over time. Organizations help ensure effective leadership by encouraging the following:

- Leadership training programs for new supervisors and refresher training programs for all supervisors
- A culture that embraces the COO system and refutes any statement or action that makes COO appear to be unimportant or classifies it as "window dressing"
- A culture that maintains a sound and predictable balance between taking corrective action and taking disciplinary action when someone makes a mistake
- Understanding among supervisors that each worker is unique. Through task assignments, verification, and other activities, effective supervisors find ways to make workers successful by leveraging their strengths and compensating for their weaknesses.
- Management support of the supervisor's role in enforcing the OD concepts
- Regular reinforcement of the safety principles that supervisors are to understand and enforce

5.5.11 Assigning Qualified Workers

Worker qualification begins at the hiring stage. Although it is beyond the scope of this book, it is important that companies have effective human resource processes to ensure that new hires (and reassigned workers) have the basic aptitude and education needed for their new position. Workers become qualified through training, experience, oversight, and corrective feedback. Workers vary in their ability to absorb certain concepts, and some are better able to perform certain skill-based tasks than others. These inherent differences should be used to the work group's advantage. In addition, specific qualifications are required in some jurisdictions for performing certain activities, such as welding.

> Workers assigned to design a twenty-inch bypass line between Reactors 4 and 6 at the Nypro plant in Flixborough, England, were competent in piping system fabrication, but not piping design. Approximately two months after the temporary jumper line was installed between Reactors 4 and 6, the bellows attached to the jumper pipe failed, leading to a large release of cyclohexane. The resulting vapor cloud explosion killed twenty-eight workers. According to the *Flixborough Report*, there was no works engineer at the time of the disaster, and no adequately qualified mechanical engineer at the site. (Ref. 5.9)

Placing highly motivated but unqualified workers in a hazardous situation is a recipe for disaster. Unqualified workers are often unaware of what they do not know and cannot see how this could be a problem. Motivated but unqualified workers too often develop a plan and proceed, giving only superficial consideration to what could go wrong. As discussed in Section 5.5.4, an effective training program culminates in a qualification step. High-performing organizations include this step and, when necessary, delay work until qualified workers are available.

5.5.12 Access Control

Separating large groups of people from hazards reduces risk. Preventing large groups of people from gathering in areas where others are doing critical work helps prevent errors.

> Shortly after 8:00 a.m. on April 16, 1947, a fire was detected in Hold 4 on the freighter *Grandcamp* while ammonium nitrate and other cargo were being loaded in Texas City, Texas. A series of poor emergency response decisions resulted in an ever-increasing fire, and the unusually colored flames drew citizens to the docks near the *Grandcamp's* slip. Clearly, a fire is not a safe condition, yet people from throughout the city moved closer to get a better look. At 9:12 a.m., more than an hour after the fire was first detected, there was an explosion that killed at least 468 people, the majority of whom were simply watching the event. (Ref. 5.24)

In the years following the *Grandcamp* explosion, bystanders have generally learned to move away from, not toward, incidents involving hazardous chemicals. However, during a process upset or highly hazardous operation, well-intentioned facility personnel often move toward the control room or other area near the process to monitor what is going on, contribute their thoughts, or be available in case an extra set of hands is needed. At best, this can become a distraction; at worst, it places more people in harm's way if things fail to go as planned. It can also create confusion, particularly if a well-intentioned but uninformed senior manager who is not charged with making emergency decisions begins to give instructions. One relatively recent development is to provide a separate gathering place for crisis management teams, emergency response teams, and operating crews who remain behind to shut down processes or take other emergency actions. With proper intergroup communication, this can help focus each team on its assignments and limit confusion.

Even when there is no special situation that draws people together, workers assigned to process units tend to gather in control rooms. If the gathering place is an unsafe location or presents a distraction to control room operators assigned to perform critical tasks, supervision must intervene.

In addition to controlling access to critical work areas, access to operating units where hazards exist should be limited to those who have the appropriate training and a need to be there. Procedures for access control should force a conversation between the entrant and the person responsible for the area (e.g., typically an operator) so that the entrant can explain why he or she needs to enter the area and what he or she plans to do while there, and the person responsible for the area can inform the entrant of any special conditions or unique hazards that exist at that time. In addition, logging the entrance and exit of all nonoperational personnel can assist with accounting for personnel in the event of an emergency evacuation.

5.5.13 Routines

We are creatures of habit. Once we determine an optimal way to drive to work, we rarely deviate from that route in the absence of specific knowledge such as a traffic report. Routines can reduce stress, and to some degree they increase safety as we become more aware of hazards (blind hills, sharp turns, etc.) along a familiar route. Likewise, turning knowledge-based activities into rule-based procedures or skill-based habits typically improves human reliability. When routines fail to follow established procedures, a trap is set, making it more likely that if other safeguards fail, an incident may occur. Routines that promote taking shortcuts or failing to conform to procedures must be detected and corrected in a timely manner.

At most facilities, there is a brief meeting each morning with key unit personnel to review results from the previous day, plan for the current day, and discuss other upcoming or unusual events. Typically, the meeting follows a set agenda, which helps ensure that all areas are addressed. However, topic order is important, and this is an excellent opportunity to reinforce values and goals. If the first agenda item is production, performance, quotas, or goals, it sends a different message than if safety performance is addressed first.

Routines promote consistency. If the intended actions become the routine, performance usually improves. However, this is not always the case. For example, when the physical arrangements of processes or units change, or if there are two similar but slightly different units in the same area, habits and routines can be erroneously applied to the wrong system. Likewise, if two similar but different procedures are needed to start up the unit under different conditions, there is an increased likelihood that workers will use the wrong procedure. When these conditions exist, special labeling and color-coding, along with greater levels of self- and peer-checking, are appropriate. However, one benefit of changing routines is to see things from a different angle. Just as outside observers often immediately spot unsafe conditions, workers on routine patrols are more likely to spot unusual situations when they are not blinded by the simple task of walking a route they cover several times each day.

Consider establishing standard routines for activities such as:

- Periodic operator patrols
- Pre-task hazard reviews
- Startup readiness reviews
- Shift turnovers
- Daily production review meetings
- Maintenance planning and coordination meetings

- Management reviews
- Capital project reviews
- Design reviews
- Hazard reviews
- Material deliveries, transfers, and shipments
- Contractor briefings
- Housekeeping

Finally, if something other than the intended action becomes the routine, performance will suffer; and just as old habits are hard to break, it will be more difficult to correct the deviation.

5.5.14 Worker Fatigue/Fitness for Duty

Fitness for duty was once a code phrase for drug and alcohol screening. While impairment due to drug or alcohol abuse does adversely affect job performance, other issues can also compromise an employee's fitness for duty. Fatigue is one common cause of impairment. Due to the high level of training required to run a process unit and the periods of high work intensity (such as turnarounds), personnel are sometimes asked to work extended shifts and to work consecutive weeks without a day off. (Specific guidelines for shift schedule, shift duration, and overtime management are included in the CCPS publication entitled *Human Factors Methods for Improving Performance in the Process Industries* and API Recommended Practice [RP] 755 [Ref. 5.25].)

On March 24, 1989, the supertanker *Exxon Valdez* ran aground on Bligh Reef in Prince William Sound, Alaska, spilling approximately 250,000 bbl of crude oil into the sea. Although there were strict rules for navigating that part of the sound, tanker captains routinely evaluated hazards posed by ice floes and oncoming traffic and intentionally steered outside of their established lane to reach open water more quickly.

On the night of the incident, the *Exxon Valdez* was maneuvering outside of the established southbound shipping lane to avoid ice. Shortly before the accident, the ship's master turned control over to the third mate, which was a violation of federal law and Exxon policy. The National Transportation Safety Board determined that one cause of the grounding was that both the ship's master and the third mate were unfit for duty at the time of the incident; the former because of alcohol consumption and the latter because of lack of rest. At the time of the incident, the blood alcohol level of the ship's master was estimated to be 0.20%, and it appears that the third mate had been on duty for most of the preceding twenty-four hours. (Ref. 5.26)

In addition to impairment and fatigue, factors that influence fitness for duty include illness, distraction due to personal issues, and mental state. Regardless of the specific cause, coworkers, supervisors, and others should be alert for signs that a worker is not fit for duty, and there should be a procedure for dealing with fitness-for-duty issues.

Things to consider when establishing procedures for dealing with fitness-for-duty issues include the following:

- Impaired workers should be moved to a safe location; if this leaves the unit understaffed, processes should be brought to a safe/stable condition promptly.
- If the need to remove a worker is not clear (e.g., the worker is tired or is suffering from a minor illness) and the worker believes that he or she can continue to work, (1) the worker's activities should be closely monitored, (2) efforts should be made to provide relief or reassign the worker to less critical tasks, and/or (3) the worker's condition should be independently evaluated (e.g., by the plant nurse or a second supervisor).
- There should be provisions for workers who do not believe that they can safely continue to work to voluntarily remove themselves from duty with administrative consequences no more severe than what would have happened if they had called in sick that day.

In addition, companies should develop policies that address fitness for duty. Issues commonly addressed in such policies include:

- Substance abuse prevention, including requirements for pre-employment, random, and for-cause drug and alcohol screening
- Protocols to be followed if a worker is believed to be impaired or otherwise unfit for duty
- Overtime, including the maximum number of hours per shift, minimum rest periods between shifts, maximum number of work days without a day off, total hours worked within a fixed period (e.g., any two-week period), and minimum number of days off prior to being eligible to return to work (a model policy is provided in API RP 755 [Ref. 5.25])
- Training for supervisors on detection of and response to personal issues (e.g., marriage issues, concerns about family members)
- Employee assistance programs to help employees with personal issues
- Employment guidelines for personnel who voluntarily enter drug or alcohol treatment programs
- Reasonable accommodation for workers with disabilities
- Return-to-work guidelines for workers who are absent or on extended medical leave

5.6 PROCESS

Processes should be (1) capable of maintaining stable conditions over the production schedule time and (2) controllable by a qualified operating team under all expected conditions and transitions. In addition, the safety margin should allow for some level of human error and machine failure (addressed in Sections 5.5 and 5.7, respectively). Furthermore, process limits and the basis for safe operation should be clearly stated; workers should not have to guess whether the process is headed out of control or is no longer safe to operate.

The attributes addressed in this section include:

1. Process capability
2. Safe operating limits
3. Limiting conditions for operation

5.6.1 Process Capability

It is unreasonable to ask workers to achieve results that are beyond the capabilities of a process. In fact, demanding that operators compensate for an unstable or incapable process is unsafe and adversely impacts COO. Faced with this challenge, well-meaning operators may begin to test various operating strategies with the intent of controlling the process better, but their ad hoc experimentation has the potential for unreliable or unsafe operation. Furthermore, human performance is variable, and it is unfair to expect the average worker to control a process that will only remain safe in the hands of an expert operator who performs flawlessly. The process should be capable of safe operation by the *least* qualified operator.

> On April 8, 1998, a runaway reaction at a Morton International, Inc. (Morton) facility in Paterson, New Jersey, resulted in a series of explosions and fires, injuring nine workers and releasing potentially hazardous materials outside of the site boundary. Investigators determined that when scaling up the process from laboratory to production scale, Morton had revised the reaction control strategy from semibatch (where one ingredient is added incrementally to minimize the potential for a runaway reaction) to full batch, where all of the ingredients are added to the reactor and the reaction rate is controlled solely by the removal of heat from the reaction mass. (Ref. 5.27)

An inherently unstable process places unusually high demands on operators to maintain control, sometimes requiring very quick decisions, precise adjustments, or unusual actions in response to upset conditions. Companies in the process industries should strive to develop processes that are fault tolerant – processes that will operate safely and under control even if one or more systems fail or the operator makes an error.

Strategies that help promote process capability include the following:

- Competent research and development personnel are consulted on process changes and, likewise, process engineers are consulted on synthesis route and other decisions that can change the inherent process hazards.
- When considering changes to the process, reviewers should constantly ask "What can go wrong?" with at least as much effort as is applied to questions such as "What are the critical success factors?"
- Systems are designed to be controllable, with safety margins sufficient to tolerate credible process variations, equipment failures, and human errors.

- Thorough design reviews are conducted that expand the question "Will this work?" to include consideration of system failures and human errors, and that ask "Is this controllable?"
- Hazard reviews are performed (separately from and after design reviews) to investigate the capability of the process to remain in a safe condition given an initiating event with the possibility of an incorrect operator response and/or failure of engineered safeguards.
- Design and hazard reviews are performed in an iterative manner (e.g., hazard reviews are not episodic; they are scheduled and performed after the design is changed).
- Hazard reviews are performed when sufficient information is available, but not so late that it is not practical to address the hazard review team's recommendations.
- Incident investigation methods consider process capability deficiencies and, when appropriate, recommend process capability improvements.
- Independent of reported incidents, facility personnel have a sense for process capability issues and, when capability issues are identified, they proactively address them.

5.6.2 Safe Operating Limits

The need for safe operating limits is well established in the process industries. Procedures should clearly state the limits, specify the actions to take to avoid exceeding the limits, and dictate the response to a process that is outside of the established safe limits. Except under extraordinary conditions, operation outside of prescribed limits should be unacceptable. This decision should not be left solely to the operator or to a collection of operational and technical personnel present at the time. The prioritization of emergency response actions should be thought through well in advance.

A facility manufactured acrylic-based powder coating and paint additives in a 1,500-gallon batch reactor that used an overhead heat exchanger to condense and return vapors to the reactor, thereby controlling the reactor temperature and preventing a runaway reaction. Upon receiving an order for slightly more product than what would be produced in a standard batch, managers elected to "scale up" the batch size and fill the order with one batch, rather than making two batches and remaining within the proven safe range of operation for the reactor and condenser. This larger batch size overwhelmed the cooling capacity of the condenser, which led to a runaway reaction that ultimately killed one worker and injured fourteen others. (At least one other COO/OD failure also contributed to this incident: Workers had adopted a practice of only tightening four of the fourteen bolts on the reactor hatch, which caused it to open and vent vapors to the production building well below the design pressure for the reactor. Although it was common practice to not tighten all fourteen bolts, this error was not identified and corrected by supervisors or managers at the facility.) (Ref. 5.28)

Even more critical is a clear understanding of the organization's core values and the knowledge of how to react to the situation given those values. For example, if the organization values process safety over output, operators will intuitively select the alternative that they believe minimizes risk, even if it is certain to result in a longer recovery time than another alternative under consideration.

The discussion about thoughtful compliance in Section 5.5.5 also applies to safe operating limits. Safe operating limits and steps to avoid or correct the deviation are often written with little or no actual operating experience at these limiting conditions. It is possible that these conditions require a somewhat different response. However, the planned actions should be followed unless they would create an imminent danger. Data from process upsets that resulted in operation outside of these limits should be reviewed to (1) assess the conditions of the plant and determine what is causing the plant to behave as it is and (2) anticipate how potential responses will affect the plant. The MOC process can be used to change the safe operating limits and procedures as necessary afterward.

Limits should be established on the basis of equipment capabilities and process dynamics. For example, if the intent is to prevent a runaway reaction, limits such as concentration of reactants, catalysts, contaminants, or inert materials might be much more useful than limits on temperature. A temperature limit that would preclude a runaway reaction given any mixture of reactants, catalysts, and inert materials might also prevent the intended reaction when the reactor contains the proper quantity and ratio of materials. Conversely, if the key variable is reactor temperature, suitable operating limits should be established based on temperature, not on another parameter that might infer temperature, such as pressure.

Limits should be based on parameters that the operator can monitor and control. For example, if the only credible cause of high temperature in a storage tank (leading to high pressure and loss of containment) is a large external fire, it would not be logical to set an operating limit for temperature, even if a temperature indicator is present. Clearly, operators cannot directly control a fire of sufficient size to significantly warm the material in a large storage tank.

When establishing safe operating limits for COO, consider the following:

- Only set operating limits for critical parameters. For example, high temperature limits are typically meaningless for liquefied gases in metal pipelines (LPG, ammonia, chlorine, etc.) because high pressure would threaten the integrity of the pipeline long before high temperature. Conversely, low pressure limits are typically meaningless for piping systems, but low temperature leading to brittle fracture can be a significant concern, regardless of the pressure.
- Only set limits within the design basis. If thermodynamics says that the process temperature cannot go below -28°F and the equipment is rated for -40°F, no safe operating limit is needed. Do not set a lower limit based on speculation that the metallurgy might be incorrect or subsequently changed. Rather, take steps to ensure that it is correct and not changed in error (see Section 5.7.4).

- Only set limits for parameters that can be measured and controlled. This does not preclude adding instruments or software logic or making changes to provide the operator with the ability to control a condition or process variable. However, it does preclude simply setting a limit and expecting the operator to figure out how to deal with it when it happens.
- Provide sufficient time for the operator to detect the condition, diagnose the situation, and take appropriate action. Limits are only useful if the operator can react in time to prevent the process from reaching an unsafe state.

5.6.3 Limiting Conditions for Operation

LCOs apply when a safety system has been deemed so important that continued operation (or at least certain activities) is prohibited when the system is not available. LCOs might include flares, scrubbers, fire-detection and suppression systems, emergency cooling, and a host of other systems that mitigate the effects of a release of process materials. Even if facilities do not normally use the term "LCO," there are instances where they are commonly applied. For example, many facilities establish an LCO for staffing levels during startup, shutdown, and ongoing operation. If minimum staffing/qualification requirements cannot be met, the activity is not started or the process is rendered safe. For example, the maximum number of personnel allowed on an offshore platform may be limited by lifeboat capacity rather than the number of beds in the accommodation block. One of the more common LCOs is that facility operation (or certain activities such as product transfers at a terminal) might be prohibited when a process flare is not in service.

Around 7:00 p.m., the rupture of a forty-two-inch water main left an entire city, including the local refinery, with no water pressure. Refinery personnel were immediately concerned about the loss of water because they depended on it for boiler feedwater makeup. Attempts to contact the city to determine when service might be restored were unsuccessful. As the level in the boiler feedwater system dropped, the decision was eventually made to shut down the refinery as rapidly as possible. This required simultaneously venting flammable gases from multiple units to the flare. When the boiler feedwater system ran dry, the boiler had to be shut down as well. Loss of steam to the flare stack resulted in structural failure. Fortunately, no one was hurt and equipment damage was limited to the flare stack.

City water was not identified as an LCO, and the operating procedures provided no guidance on how to respond to its loss. The team on shift at the time of the incident knew that steam was used to improve the flare's combustion efficiency, but they were unaware that steam was needed to cool the stack under peak flaring conditions. Also, the team did not know how quickly they needed to begin shutting down the refinery in order to avoid losing steam flow during the shutdown. Under the circumstances, the crew that evening did a good job of safely shutting down and minimizing equipment damage. However, it was fortunate that the supervisor on shift made the call to shut down the refinery while there was still a significant amount of boiler feedwater available.

Incidents such as Piper Alpha (discussed in Chapter 3) and the Isomerization Unit (discussed in Chapter 1) demonstrate that workers sometimes fail to shut down processes, even in the face of extreme adversity. Most personnel in the process industries are skilled problem solvers, and shutting down a process is a clear admission that the problem was not solved. Too often, the heroes in the facility are the ones who gamble that they can successfully navigate under unsafe conditions and avert a costly outage. However heroic, accepting intolerable risk is poisonous to a COO/OD system. Like safe operating limits, LCOs make it clear when the shift team should stop troubleshooting and execute the shutdown procedure.

Some LCOs apply to nonroutine activities. For example: (1) hot work is not authorized when sprinklers and other fire-protection systems are out of service, (2) confined space entry is not permitted unless there is a sufficient number of trained emergency response personnel on site to conduct a confined space rescue, and (3) nonessential personnel are not allowed in a unit, or adjacent units, during startup. These conditions should be included in written procedures governing the nonroutine activity and on checklists used to authorize the activity.

Facilities should carefully review their basis for safety and consider establishing LCOs based on any or all of the following:

- Availability of safety systems such as flares, scrubbers, and fire-protection systems
- Failure of offsite or onsite utility systems
- Availability of backup systems necessary for safe shutdown in the event of a failure of the primary system
- Adequate staffing
- Special activities or operating modes

5.7 PLANT

Maintaining equipment to be fit for service is as important as the attention paid to the people and process aspects of COO. However, COO goes beyond fitness for service – it helps ensure that there is a clear “owner” at all times. In addition, it sets out standards for monitoring and controlling equipment.

The attributes addressed in this section include:

1. Asset ownership/control of equipment
2. Equipment monitoring
3. Condition verification
4. Management of subtle changes
5. Control of maintenance work
6. Maintaining the capability of safety systems
7. Controlling intentional bypasses and impairments

5.7.1 Asset Ownership/Control of Equipment

There should never be any doubt about who is in control of each asset and the land surrounding process units. Whenever control of an asset passes from one individual to another, there should be a structured handoff to ensure continuity.

When asked who owns an asset, personnel at almost all facilities reply that the group assigned to operate the asset is the owner. This makes sense, and it closely parallels how we think about our personal property. An auto mechanic does not "own" our car simply because we dropped it off for repairs or maintenance. However, while the car is at the shop, the mechanic is in control of the asset (within boundaries established by the owner). For example, it is appropriate for the mechanic to test drive the car, but driving the car for any other purpose is not appropriate. Likewise, clear rules should establish who "owns" each asset, and there should be limits or conditions imposed on activities performed by personnel who are not the "owner."

A fill-in operator was running a packaging line at a food processing facility. The line jammed, requiring the operator to stop the line and clear the jam. An operator assigned to an adjacent line noticed that the line was not running. Thinking that the fill-in operator had stopped the line to take a break, the permanent operator on the adjacent line removed the stop tag and restarted the line. Fortunately, the fill-in operator noticed some movement and quickly removed his hands from the machine, preventing a likely amputation. Although this was a clear violation of the facility's lockout policy, it should never have happened with or without the machine tagged out. The operator on the adjacent line was at no time in control of the line that was shut down. Because of that, he had no authority to start it up, regardless of the experience level of the person assigned to run the line.

The error described above, where one operator removed a tag and started equipment without permission, is a very rare event in the process industries. However, that is not because it cannot occur; rather, it is because facilities normally have well-established rules that prohibit maintenance, technical, and other support personnel from operating process equipment. (The exception might be utility systems where the maintenance department is the system's owner/operator.)

Ownership changes over the life of an asset. During initial construction, the project manager and project team typically are responsible for the asset. Regardless of how badly operations or maintenance wants a pump moved to a more accessible location, it does not get moved until the project manager approves the change. At some point, there is a defined handoff to the commissioning team (which may be a subset of the project team). Another formal handoff occurs between the commissioning team and the operating group. From that point, standard management systems (such as work permits) govern the handoff between operations and maintenance, and other systems govern the handoff to and from the turnaround management team. Finally, there should be a clear protocol for transfer of control during an emergency, and a protocol within the emergency response program for transfer of incident command.

A common point of contention is the authority to make equipment available for scheduled maintenance. If operations is solely accountable for output and maintenance is accountable for availability, tension will be created when operations refuses to shut down equipment for preventive maintenance. One proven COO method to address this conflict is to implement a more balanced scorecard, holding both operations and maintenance accountable for availability and conformance to planned maintenance schedules.

All equipment should be "owned" by someone who is responsible for monitoring it, verifying its condition, controlling changes, and ensuring that it is properly maintained. This is particularly important for safety systems and associated equipment; these critical systems cannot be assigned to "everyone" to look after because too often "everyone" decides it is "someone else's" responsibility and "nobody" assumes any ownership.

5.7.2 Equipment Monitoring

Fundamentally, there are two maintenance strategies: (1) reactive maintenance – fixing things when they break – and (2) proactive maintenance – planning and conducting a set of activities to help prevent failures or detect the onset of failure, allowing for smooth transitions to planned repairs. For a successful COO system, the latter option is the logical choice for most components. However, regardless of the maintenance strategy, monitoring process conditions and equipment is an important duty within COO.

Early in the nineteenth century, steam-powered machines promised to revolutionize life. By mid-century, steam powered everything, from locomotives to ships to industrial equipment. However, as the complexity and power of steam engines increased, so did the frequency of boiler explosions. During the 1850s boiler explosions were occurring at the rate of one almost every four days, and in 1865 a single boiler disaster aboard a river boat killed 1,200 people.

As a result of the safety concerns related to steam power, several businessmen located in Hartford, Connecticut, formed the Polytechnic Club. This group did not agree with the popular consensus of the day – that boiler explosions were "Acts of God"; rather, they believed that such potential explosions could be detected and prevented by proper design and regular boiler inspections. In 1866 members of the Polytechnic Club founded The Hartford Steam Boiler and Inspection Company, and they provided financial incentives in their insurance products for clients who ensured that their boilers were designed properly, inspected their boilers regularly, and maintained boilers fit for service. The practice of boiler inspections proved to be effective and was widely adopted; it was eventually mandated throughout most of the United States. (Ref. 5.29)

A manager in a chemical plant once noted that an average operator appears to be too relaxed 80% of the time and too busy 20% of the time, and that the really good operator appears to be too relaxed 95% of the time. There is more than a little truth to this observation. A successful operator diligently monitors process equipment and conditions, detects and investigates unusual conditions, diagnoses the cause, and responds to the issue quickly enough to make the job look easy. Conversely, an operator who fails to monitor process equipment and conditions can make the job seem very difficult.

Routine patrols using reading sheets that include acceptable ranges for each process parameter help ensure that operators monitor equipment and help show any unusual trends. Well-developed patrol sheets can also help ensure that operators walk though each part of the process, providing the opportunity several times each day to detect unusual sounds, vibrations, odors, or other signs that something is amiss. Some facilities use commercially available handheld electronic devices to help operators collect data from field instruments and devices so that the data can be uploaded and compared to information being captured by process control systems.

Equipment monitoring cannot always be left solely to the field operator's five senses. Some critical operating parameters, such as vibration and small flange and packing leaks, can only be reliably detected with electronic sensors. Routine patrols should be supplemented by condition monitoring and other tests. Many facilities expand the operator's duties to include monitoring equipment parameters (such as levels in lubrication oil reservoirs), monitoring utility systems, checking the status of safety equipment, and collecting data to support the vibration analysis program.

In addition, a change in performance, even within what is generally considered an acceptable range, is often noteworthy. Equipment logs (or similar data from a computerized maintenance management system) can provide early signs of failure. For some parameters, such as thickness or vibration, records are invaluable because the rate of change can be just as significant as the current condition. In addition, repair logs can highlight trends that might be due to a subtle change in the quality of repair parts or the effectiveness of training.

Effective equipment monitoring programs have the following traits:

- Responsibilities are clearly assigned for collecting and analyzing data.
- Data collected manually is documented on a form that includes the acceptable/normal range, and standing procedures are in place for reporting unexpected or out-of-range results.
- Where appropriate, data are collected using electronic sensors, and the responsibility for reviewing these data is clearly assigned.
- Personnel assigned to routine patrols understand that they are to report and/or correct any nonconforming conditions, such as missing or loose pipe hangers and junction box covers, plugs missing from bleed or drain lines, missing or unreadable labels, etc.

5.7.3 Condition Verification

Condition verification is often a critical step in any procedure. When changing oil in a car, it would be unacceptable to simply add a specified quantity of oil without confirming the level by checking the dip stick. Most maintenance tasks in the process industries are much more complex than changing engine oil, yet far too often the verification steps and pass/fail criteria are not specified.

If one understands that errors can lead to unacceptable consequences, verification becomes the norm. However, if human errors and the resulting incidents are deemed unavoidable, the facility will sustain a much higher number of incidents. Either condition becomes a positive feedback loop, in the former case leading to lower incident rates and in the latter case leading to higher incident rates. High-performing facilities embrace verification and find ways to make it more systematic and effective.

An electrical failure required that a pump's motor starter and start/stop circuitry be replaced. Once the work was completed, operators used field switches to verify that the pump started and stopped correctly. However, the remote start/stop functionality was not tested. A short time later, the pump was placed in service without incident. Several weeks later, an unrelated process issue required that the unit be briefly shut down. The board operator remotely stopped the pump and closed the suction and discharge valves. A few minutes later, the operator silenced a high temperature alarm associated with the pump and took no further action because he knew he had stopped the pump. The status indicator on the control system showed that the pump was off. While the operator was busy attending to the process upset that caused the temporary outage, the overheated pump exploded. The investigation team determined that during a recent repair, the pump had been incorrectly wired such that it could not be stopped remotely.

Verification activities should be considered for:

- Critical operational tasks, particularly if failure to properly perform the task might directly lead to an incident
- Post-maintenance activities, with particular emphasis on critical functions/systems, even if there was no intent to modify those functions
- Tasks that, when performed properly, become the primary basis for safety for a subsequent activity or mode of operation
- Repeated alarms, even if other indications appear to be normal

5.7.4 Management of Subtle Changes

Ongoing operations often introduce subtle changes. Suppliers are continuously improving their parts and materials or sometimes making changes to reduce costs that actually degrade quality. Maintenance teams strive to make plant equipment more reliable, maintainable, and operable. In addition, operating units are under increasing pressure to control costs, resulting in pressure to reduce production cycle times; increase the intervals between planned maintenance activities; discontinue nonproductive activities; and use less expensive raw materials, repair parts, and other supplies.

To reduce moisture-induced clumping (which caused process problems), a supplier modified its packaging from simple paper bags to plastic-lined paper bags. This solved the clumping problem, but it eventually led to a flash fire inside a tank at a customer's facility due to static discharge as the operator added the dry powdered material via the manway.

Engineers, maintenance employees, contractors, and others who specify production and maintenance materials and spare parts and who determine operating and maintenance practices should be keenly aware of the impact of subtle changes. What appear to be improvements can increase the risk of equipment failure or other process hazards.

Features that help prevent introduction of unsafe subtle changes include the following:

- Designation of critical parts, supplies, and raw materials, where no changes (including supplier changes) are permitted without careful, documented review
- Systematic maintenance and periodic review of engineering standards and specifications to disallow materials or designs that have been found to be less than adequate
- Understanding among purchasing agents of specifications and standards, and intolerance of substitution unless specific conditions are met (e.g., equivalent specifications, end user approval)
- Keen awareness among all maintenance employees of the potential impacts of subtle changes, and a tendency to classify anything different (size, shape, appearance, form, quantity, action, etc.) as a change rather than replacement-in-kind
- Keen awareness among all employees that subtle changes may originate outside the facility (e.g., unusually hot/cold weather, airborne dust, emissions from a neighboring facility, new neighbors)
- Thorough evaluation of the implementation of changes that cannot be readily observed (such as changes to computer systems) that goes beyond basic testing to see if it "works" under normal conditions

5.7.5 Control of Maintenance Work

Maintenance work should be controlled at multiple levels. First, when someone enters a work request, it should be reviewed to ensure that it (1) is an authorized change, (2) is something that should be done, and (3) conforms to the facility's operating and maintenance strategy. Second, maintenance work should be planned and coordinated with the maintenance and production groups. Good planning is an essential part of an effective maintenance strategy, and adherence to the plan is a good indicator of the effectiveness of a facility's COO system. Third, all maintenance work in the process area should be authorized by the team that is responsible for the area.

A factory service engineer was tasked with fixing instability in a power-generating unit. The engineer tried various adjustments to control the ramp-up rate, temperature limits, speed limits, and other parameters. Each failed restart resulted in more unburned gas reaching the exhaust system. When the problem was eventually solved and the turbine lit off, the resulting explosion destroyed the power turbine, the gas turbine, the enclosure, and the exhaust ducting bellows. The incident investigation team later determined that the factory service engineer was not authorized or qualified to execute a series of "on-the-fly" software modifications.

The first attribute presented in this section concerned ownership of equipment. Assignment of ownership is fundamental and a key element in controlling work. Owners authorize work and are most often accountable for (1) preparing equipment for work and (2) ensuring that it is not returned to service until the work is complete. However, in practice this becomes a joint responsibility, with independent overlapping safeguards embedded in operating, safe work, and maintenance procedures.

In addition to clear assignment of responsibilities for controlling work and effective safe work practices, activities that help promote control of maintenance work include:

- Review of all work requests by qualified personnel to ensure that the requested work does not require change authorization (or authorization has been granted), the work plan is technically sound, and the work plan conforms to facility standards
- Written work permit authorizations that include a description of the work and the hazards, safety requirements, and any required communication steps or hold points
- A practice to periodically audit the effectiveness of the work permit system and routinely inspect work in progress to determine whether work in the field conforms to permit conditions
- Separate departmental locks applied by (1) the equipment owner and (2) the maintenance crew until it is verified by authorized persons in each department that the work is complete and the equipment is safe to operate
- Written procedures that help ensure that work is properly completed, including steps such as checking for leaks, checking rotation, aligning equipment, testing any safety systems that may have been affected by the work, etc.

5.7.6 Maintaining the Capacity of Safety Systems

Almost all facilities know when they lose the capability to produce product. Closely tracked production or product quality numbers dip, and the business team reacts quickly. However, failure to maintain the capability of backup or safety systems can go undetected for months or years. Operating with insufficient capability, whether in terms of safety systems, backup systems, or some other aspects of the operation, at best causes unreliable operation and at worst causes the final failure in a chain of events that lead to disaster.

Designers often provide high levels of redundancy in critical utility and safety systems to increase system reliability. Too often, this intent is defeated when these systems are not maintained. In other cases, designers fail to provide sufficient backup systems, or backup systems that appear to be redundant have an undetected common cause failure mode. (Design flaws often result from COO failures on the part of engineering, such as failure to perform thorough design and hazard reviews.) When safety systems fail, operators can be forced to quickly make tough decisions. A rapid decision to shut down or take alternate actions, typically based on uncertain or incomplete information, cannot be optimal and is too often wrong.

Failure of firefighting systems in the aftermath of the 1989 fire and explosion at the Phillips Houston Chemical Complex in Pasadena, Texas, contributed to the magnitude of the incident. Process water and firefighting water shared the same systems, and the initial blast caused extensive damage to this system as well as the electrical distribution system needed to power the main pumps. However, there were three diesel-driven firewater pumps. Unfortunately, one pump was out of service for maintenance, the engine that drove the second pump ran out of fuel within an hour (due to failure to maintain the intended level in the fuel tank), and the third pump failed during the course of the emergency response activities. Failure of these pumps provides a clear demonstration of how critical safety systems might suffer in the absence of proper maintenance and testing.

For example, analyses of the Piper Alpha incident (see Chapter 3) often focus on the failure of the permit-to-work system that allowed Condensate Pump A to be returned to service even though the relief valve on Pump A had been replaced with a thin metal plate. Certainly, this was a major failure. However, Pump A was expected to be out of service for up to two weeks, leaving the platform vulnerable to unplanned shutdown if Pump B went down for more than a few minutes. If the mean time between twenty-minute failures of a pump falls between one and ten years, the likelihood of such a failure during a two-week period is between 0.4% and 4%. In process safety terms, those are very high initiating event rates, particularly when operators have no backup plan. The greater the likelihood of failure for important utility or safety systems, the more important it is to provide an alternate or backup system or to develop and practice a procedure for how to react to the failure. Clearly, the only two options considered on the Piper Alpha platform on the evening of July 6, 1988, were shutting down the platform or restoring Pump A to service. Under severe time pressure, platform personnel made the wrong choice.

Maintaining capability requires:

- Fault-tolerant designs
- Diligent maintenance, testing, and repair of backup systems
- Effective contingency plans for when equipment does fail

5.7.7 Controlling Intentional Bypasses and Impairments

Almost all process units are fitted with numerous safety systems, many of which involve interlocks that take the process to a safe state when some sort of fault or unsafe condition is detected. At times, these systems trip spuriously, leaving the facility to decide whether to impair or bypass the system, or to leave the process down pending repair. In many cases, and for a variety of good reasons, these systems are bypassed for a period of time.

It should not be easy to override or impair safety systems. These activities should require formal written requests for a temporary change; authorization for a short period of time; and, preferably, special tools, keys, or passwords for execution. If a system can be easily overridden by the team assigned to operate the facility, the COO/OD system will be challenged to ensure that overrides do not become a common practice.

> On April 23, 2004, a large release of flammable materials occurred at the Formosa Plastics Corp. facility in Illiopolis, Illinois, when an operator intentionally forced the drain valve open on a reactor containing vinyl chloride monomer (VCM). The capability to locally override the interlock that held the bottom valve closed during the reaction phase was installed several years earlier to allow operators to manually transfer part of the contents of a reactor into a second empty reactor in the event that the reaction rate could not be controlled by other means. (To override the interlock, the operator simply connected an air hose to the bottom valve, forcing it open.) Although the facility's policy prevented using this local override except when authorized by the supervisor, it appears that the operator assigned to drain wash water from Reactor 306 mistakenly forced open the drain valve on Reactor 310 without authorization, apparently assuming that there was some sort of fault in the control system. The VCM release resulted in a fire and explosion, killing five people and injuring three others at the facility. (Ref. 5.30)

Overrides or impairments can take many forms. Historically, it was often a "jumper" wire in parallel with an electrical relay or switch that trips the interlock when the circuit is open or de-energized. However, interlocks configured in programmable systems can normally be impaired in any number of ways via software changes. To allow for online testing, many shutdown systems are fitted with bypass lines around valves that trip to the closed position. Simply opening the bypass line will disable the interlock. Some pressure relief valves are fitted with block valves on the inlet and outlet of the device to allow for testing between turnarounds, and the relief system can be impaired simply by closing either of the block valves.

Although there are many ways to impair or bypass safety systems, there are fundamentally four types of safeguards that help ensure the dependability of these types of systems; effective COO systems employ all four:

1. All impairments or overrides of safety systems require authorization, and the person authorizing the activity considers both (a) the alternate

safeguards that are in place or proposed and (b) the current conditions and pace of operations for the unit.

2. The means to impair the system requires some form of special tool, special knowledge, key, or other access designed to require that at least two people collaborate on the decision to impair the system.
3. Impairment of a safety system triggers an automatic "alarm clock" that either (a) returns the system to functionality after a specified time, (b) requires someone to confirm that the system has been made functional, (c) requires reauthorization, or (d) provides an equivalent means to force a decision to remove or reauthorize the impairment.
4. There is a requirement to check periodically the status of systems that can be impaired accidentally or without special tools or knowledge (e.g., a tamper-indicating seal on a manual valve under a relief valve). The time period should correspond to the likelihood of unintended impairment, and the rules for impairment should conform to applicable codes and standards.

In addition, one other method that helps promote prompt repairs to safety systems is periodic reporting of the number of authorized impairments or overrides as part of the metrics. Management's typical response to this is "Why is the number not zero?" The resulting discussion often increases the priority of work orders to repair these systems, thus reducing the total time the safety system is unavailable.

5.8 MANAGEMENT SYSTEMS

This section addresses management systems that are (1) related to COO but not within the scope of this book and (2) necessary conditions for an effective COO system that are fully addressed in other CCPS guidelines. These management systems include:

- Related programs
  - standards of conduct
  - evaluation/performance assurance
- Necessary conditions
  - hazard evaluation
  - safe work practices
  - management of change
  - planning for/responding to emergencies
  - audits, inspections, and critiques

5.8.1 Related Programs

It becomes difficult to separate COO from organizational culture, and the CCPS continues to emphasize this topic (Ref. 5.31). It is no coincidence that process safety culture was included as the first process safety element in the CCPS's *Guidelines for Risk Based Process Safety* (Ref. 5.2). Essential features of developing and maintaining a sound process safety culture (Refs. 5.2 and 5.32) include:

- Espousing safety as a core value
- Providing strong leadership
- Establishing and enforcing high standards of performance
- Maintaining a sense of vulnerability
- Empowering individuals to successfully fulfill their safety responsibilities
- Providing deference to expertise
- Ensuring open and effective communications
- Establishing a questioning/learning environment
- Fostering mutual trust
- Providing timely response to safety issues and concerns
- Providing continuous monitoring of performance

Clearly, there is significant overlap between COO and process safety culture, particularly in the areas of leadership, empowerment, communications, learning, timely response, and monitoring of performance. A culture that embraces "doing what we say we will do" is a good candidate for the next step – COO. On the other hand, an organization that places great value in "doing what nobody else has done before" and, in particular, embraces a very high level of individual creativity will have difficulty implementing a COO system. There will be too much conflict between COO discipline and the organization's culture. Without significant effort, any new initiative that is in conflict with an organization's culture is likely to fail.

Section 5.5.9 points out the importance of verification and presents it within the context of task verification. Verification should be extended to management systems and programs. High-performing organizations establish meaningful metrics, including metrics that help predict the future (see Section 7.4.1). For example, a high level of compliance with procedures is an indicator of lower risk, assuming, of course, that the procedures accurately describe how to do the task. COO provides many opportunities to measure conformance to a standard, and tracking this should help (1) provide early warning of increasing risk and (2) monitor the COO system.

Other metrics might include the following:

- Percent of incidents where failure to follow procedures or inadequate training is listed as a root cause
- Percent of qualified personnel in defined roles
- Staff turnover rates
- Number of incidents attributed to trainees
- Number of nonroutine and emergency maintenance work orders
- Number of audit findings related to inoperable instruments and tools

- Number of housekeeping audits and their scores
- Number of incidents caused by a lack of self-checking or peer-checking
- Percentage of overtime hours
- Number of unplanned shutdowns
- Number of unplanned safety system activations
- Number of unplanned safety system activations for invalid reasons

5.8.2 Necessary Conditions

Clearly, the margin of safety at a facility quickly erodes once management stops taking an interest in the conditions and programs necessary to support safe operation. The need for effective PSM systems was highlighted by the Bhopal incident and several others that followed in the 1980s. The CCPS published its first comprehensive book on PSM systems in 1989 and followed it up with a second edition three years later (Refs. 5.33 and 5.34). In 2007, the CCPS updated and expanded its guidance with the publication of *Guidelines for Risk Based Process Safety* (Ref. 5.2).

At some level, effective PSM systems and COO are inseparable. Certainly, some aspects can be done well without a comprehensive COO system. For example, a process hazard analysis (PHA) team might do a very good job of identifying hazards. However, without the management commitment to resolve the PHA team's recommendations in a timely and effective manner, little good comes from the effort. The latter part, addressing PHA recommendations in a timely and effective manner, cannot be separated from COO. In fact, auditors typically observe that facilities that address PHA recommendations effectively also tend to address incident investigation team recommendations and audit findings effectively, even when different people are assigned to the three elements. The reason why these and many other PSM activities are effective is a strong commitment to maintaining sound management systems, a vital step in developing an effective COO system.

The fact that COO/OD systems can complement, but cannot replace, proven PSM systems cannot be overstated. For example, without sufficient process safety knowledge, a diligent PHA team might do a very good job of adhering to the meeting schedule and documenting the results of the analysis, and the facility might do a very good job of resolving recommendations and addressing action items. However, with insufficient process safety knowledge, the risk associated with analysis uncertainty (i.e., scenarios not postulated and analyzed by the team) can overwhelm the risk benefit provided by addressing the team's recommendations.

Organizations should not be satisfied with proven adherence to policies, procedures, and practices. Since COO generally does not promote creativity and, to a degree, may limit people's tendency to question why things are done in a certain manner, it is very important to strike a balance between high levels of creativity (which within the process safety context might be needed to identify hazards or improve methods) and high levels of conformance. Overemphasizing either aspect can increase risk.

5.9 SUMMARY

This chapter addressed several key attributes of COO (dividing them into attributes that span an entire organization and attributes that relate primarily to people, process, and plant) and the relationship between PSM systems and COO. Adopting each attribute should be beneficial, but some will have a much greater risk benefit than others. Moreover, not all of the attributes will apply to any given facility.

Readers should review each attribute listed in Table 5.1 and consider the COO gaps that might exist at their facility. Armed with that information, along with the descriptions of the COO attributes in this chapter and the OD attributes in Chapter 6, readers should determine improvement objectives and take steps to establish a new COO system or improve an existing one.

TABLE 5.1. Summary of COO Attributes

Foundations	1. Understand risk significance 2. Establish standards that support the organization's mission and goals 3. Understand what can be directly controlled and what can only be influenced 4. Provide the resources and time necessary to complete tasks within standards 5. Ensure competency across the organization 6. Perform critiques and take corrective action
People	1. Clear authority/accountability 2. Communications 3. Logs and records 4. Training, skill maintenance, and individual competence 5. Compliance with policies and procedures 6. Safe and productive work environments 7. Aids to operation – the visible plant 8. Intolerance of deviations 9. Task verification 10. Supervision/support 11. Assigning qualified workers 12. Access control 13. Routines 14. Worker fatigue/fitness for duty
Process	1. Process capability 2. Safe operating limits 3. Limiting conditions for operation
Plant	1. Asset ownership/control of equipment 2. Equipment monitoring 3. Condition verification 4. Management of subtle changes 5. Control of maintenance work 6. Maintaining the capability of safety systems 7. Controlling intentional bypasses and impairments

Chapter 3 described the management leadership and commitment that are the foundation of the COO system, and this chapter provided a menu of proven activities that improve organizational performance. However, these improvements do not happen until members of the organization change their day-to-day behavior. Chapter 6 explores ways in which people fail to perform in a prescribed manner and offers suggestions for addressing these issues. Chapter 7 describes methods for implementing a COO/OD system.

5.10 REFERENCES

5.1. Kletz, Trevor, *What Went Wrong? Case Histories of Process Plant Disasters, Fourth Edition*, Elsevier, Burlington, Massachusetts, 1999.

5.2. Center for Chemical Process Safety of the American Institute of Chemical Engineers, *Guidelines for Risk Based Process Safety*, John Wiley & Sons, Inc., Hoboken, New Jersey, 2007.

5.3. American Chemistry Council, *Responsible Care® Management Systems and Certification,* http://www.americanchemistry.com/s_responsiblecare/doc.asp?CID=1298&DID=5086.

5.4. API Recommended Practice 75, *Development of a Safety and Environmental Management Program for Offshore Operations and Facilities, Third Edition*, American Petroleum Institute, Washington, D.C., May 2004.

5.5. U.S. Department of Energy, DOE Order 5480.19, Change 2, *Conduct of Operations Requirements for DOE Facilities*, Washington, D.C., October 23, 2001.

5.6. Atherton, John, and Frederic Gil, *Incidents That Define Process Safety*, Center for Chemical Process Safety of the American Institute of Chemical Engineers, John Wiley & Sons, Inc., Hoboken, New Jersey, 2008.

5.7. Kletz, Trevor, *Lessons from Disaster: How Organizations Have No Memory and Accidents Recur*, Gulf Publishing Company, Houston, Texas, 1993.

5.8. Kletz, Trevor, *Still Going Wrong! Case Histories of Process Plant Disasters and How They Could Have Been Avoided*, Butterworth-Heinemann, Burlington, Massachusetts, 2003.

5.9. Lees, Frank P., *Loss Prevention in the Process Industries: Hazard Identification, Assessment and Control, Second Edition*, Butterworth-Heinemann, Oxford, England, 1996.

5.10. Rogers Commission, *Report of the Presidential Commission on the Space Shuttle Challenger Accident*, Washington, D.C., June 6, 1986.

5.11. Center for Chemical Process Safety of the American Institute of Chemical Engineers, *Guidelines for Hazard Evaluation Procedures: Second Edition with Worked Examples*, John Wiley & Sons, Inc., Hoboken, New Jersey, 1992.

5.12. Covey, Steven R., *The 7 Habits of Highly Effective People: Powerful Lessons in Personal Change*, Simon & Schuster, New York, New York, 1990.

5.13. U.S. National Transportation Safety Board, *Marine Accident Brief*, NTSB/MAB-05/01, Accident No. DCA-01-MM-022, Washington, D.C., 2001.

5.14. Hopkins, Andrew, *Lessons from Longford: The Esso Gas Plant Explosion,* CCH Australia Limited, Sydney, Australia, 2000.

5.15. U.S. Chemical Safety and Hazard Investigation Board, *The Explosion at Concept Sciences: Hazards of Hydroxylamine*, Case Study No. 1999-13-C-PA, Washington, D.C., March 2002.

5.16. Garvin, David A., *Learning in Action: A Guide to Putting the Learning Organization to Work*, Harvard Business School Press, Boston, Massachusetts, 2000.

5.17. National Aeronautics and Space Administration, *Mars Climate Orbiter Mishap Investigation Board Phase I Report*, Washington, D.C., November 10, 1999.

5.18. Civil Aviation Authority Safety Regulation Group, *Flight Crew Training: Cockpit Resource Management (CRM) and Line-Oriented Flight Training (LOFT)*, CAP 720, West Sussex, England, August 1, 2002.

5.19. Swain, Alan D., *Design Techniques for Improving Human Performance in Production*, A. D. Swain, Albuquerque, New Mexico, 1986.

5.20. U.S. Chemical Safety and Hazard Investigation Board, *Investigation Report: Little General Store – Propane Explosion (Four Killed, Six Injured)*, Report No. 2007-04-I-WV, Washington, D.C., September 2008.

5.21. U.S. Chemical Safety and Hazard Investigation Board, *Investigation Report: Combustible Dust Hazard Study*, Report No. 2006-H-1, Washington, D.C., November 2006.

5.22. National Aeronautics and Space Administration, *Columbia Accident Investigation Board, Report Volume 1*, Washington, D.C., August 2003.

5.23. Haynes, Alex B., M.D., M.P.H., et al., "A Surgical Safety Checklist to Reduce Morbidity and Mortality in a Global Population," *The New England Journal of Medicine*, Massachusetts Medical Society, Waltham, Massachusetts, Vol. 360, No. 5, January 29, 2009, pp. 490-499.

5.24. Stephens, Hugh W., *The Texas City Disaster, 1947*, University of Texas Press, Austin, Texas, 1997.

5.25. ANSI/API Recommended Practice 755, *Fatigue Risk Management Systems for Personnel in the Refining and Petrochemical Industries*, American Petroleum Institute, Washington, D.C., April 2010.

5.26. Howlett, H. C., II, *The Industrial Operator's Handbook: A Systematic Approach to Industrial Operations, Second Edition*, Techstar, Pocatello, Idaho, 2001.

5.27. U.S. Chemical Safety and Hazard Investigation Board, *Investigation Report: Chemical Manufacturing Incident (9 Injured)*, Report No. 1998-06-I-NJ, Washington, D.C., August 16, 2000.

5.28. U.S. Chemical Safety and Hazard Investigation Board, *Case Study: Runaway Chemical Reaction and Vapor Cloud Explosion (Worker Killed, 14 Injured)*, Report No. 2006-04-I-NC, Washington, D.C., July 31, 2007.

5.29. Hartford Steam Boiler Inspection and Insurance Company Web site, http://www.hsb.com/about.asp?id=50.

5.30. U.S. Chemical Safety and Hazard Investigation Board, *Investigation Report: Vinyl Chloride Monomer Explosion (5 Dead, 3 Injured, and Community Evacuated)*, Report No. 2004-10-I-IL, Washington, D.C., March 2007.

5.31. Center for Chemical Process Safety of the American Institute of Chemical Engineers, *Building Process Safety Culture: Tools to Enhance Process Safety Performance*, New York, New York, 2005.

5.32. Center for Chemical Process Safety of the American Institute of Chemical Engineers, *Safety Culture: What Is At Stake*, New York, New York.

5.33. American Institute of Chemical Engineers, *Guidelines for Technical Management of Chemical Process Safety,* New York, New York, 1989.

5.34. American Institute of Chemical Engineers, *Plant Guidelines for Technical Management of Chemical Process* Safety, New York, New York, 1992 and 1995.

5.11 ADDITIONAL READING

- Center for Chemical Process Safety of the American Institute of Chemical Engineers, Daniel A. Crowl, ed., *Human Factors Methods for Improving Performance in the Process Industries*, New York, New York, 2007.
- Davis, Lee, *Man-Made Catastrophes: From the Burning of Rome to the Lockerbie Crash*, Facts on File, Inc., New York, New York, 1993.
- Klein, James A., "Two Centuries of Process Safety at DuPont," *Process Safety Progress*, American Institute of Chemical Engineers, New York, New York, Vol. 28, Issue 2, June 2009, pp. 114-122.

6
KEY ATTRIBUTES OF OPERATIONAL DISCIPLINE

6.1 INTRODUCTION

Operational discipline, as defined in Section 1.4, is simply the performance of all tasks correctly every time. An effective OD system leads to very predictable behavior and actions that closely conform to the actions that have been prescribed.

- OD is the execution of the COO system by individuals within the organization.
- OD refers to the day-to-day activities carried out by all personnel.
- Individuals demonstrate their commitment to process safety through OD.
- Good OD results in performing the task the right way every time.
- Under an effective OD system, personnel recognize unexpected situations and respond to them by maintaining (or putting) the process in a safe configuration while seeking wider expertise to ensure personal and process safety.

OD complements COO, which encompasses supporting management systems that are developed, implemented, and maintained in an effort to (1) perform operational tasks in a deliberate and structured manner consistent with underlying risk assessments, (2) ensure that each task is performed correctly, and (3) minimize variations in performance.

> In almost any organization, there are some work groups that perform work activities in a structured, highly disciplined, and predictable manner while other groups flounder. Isolated pockets of excellence are not a sign of an effective OD system; more likely, they are due to effective individual leadership and positive work group dynamics. A properly functioning OD program is effective throughout the entire organization.

Successful implementation of the COO attributes described in Chapter 5 underpins the OD system. A COO system provides structure for people working together. The OD system, as described in this chapter, focuses much more on group and individual attributes that determine behavior. In short, it focuses on how people work, in groups and individually.

DuPont first included OD in its PSM program in 1989. For more than twenty years, the graphic shown in Figure 6.1 has defined DuPont's PSM program. The rim of the wheel, labeled "Achieving 'Operational Excellence' Through Operational Discipline," is supported by the

Engineer Violates Work Rules, Resulting in Commuter Train Failing to Stop

On September 12, 2008, a commuter train collided head on with a Union Pacific freight train near Los Angeles, California. Twenty-five people were killed and 101 were injured when the locomotive on the commuter train was shoved 50 feet backward into the first passenger car. The commuter train was routed and signaled to stop on a siding approximately 0.3 mile west of the point of contact to wait for the freight train to pass on the single main-track section. However, the commuter train proceeded through the siding and back onto the main track, striking the freight train less than a minute later.

Records indicate that in the eighty minutes immediately preceding the collision, the commuter train engineer used his personal cell phone to send and receive several text messages and to make two outgoing phone calls (including sending one text message seconds before the collision). Furthermore, a review of the engineer's text messages indicated that earlier in the day, he had been in communication with another person (who was neither an employee of the commuter line nor a qualified train operator) about plans to allow this second, untrained person to operate the train later that evening. Other cell phone records indicated that the same engineer had allowed at least two unauthorized persons to "ride along" in the cab just days before the fatal crash. (Ref. 6.1)

spokes, yet it provides rigidity for the entire structure. DuPont's proven OD system focuses on organizational and individual characteristics as follows (Ref. 6.2):

- Organizational characteristics
  - Leadership focus
  - Employee involvement
  - Practice consistent with procedures
  - Excellent housekeeping
- Individual characteristics
  - Knowledge
  - Commitment
  - Awareness

DuPont and many other CCPS member companies have found that an effective OD system leads to predictable behaviors and reliable human performance. In other words, the behavioral outcome is the primary measure of OD effectiveness, and vice versa. While OD focuses on personal commitment and behavior, COO focuses on management systems for specific activities, such as issuing safe work permits (SWPs). Leading and lagging indicators can be used to assess the effectiveness of the COO and OD aspects of management systems, such as the indicators for safe work practices shown in Table 6.1.

The OD system is particularly well suited for the use of leading indicators. OD metrics might include the following:

- Percentage of incident investigations that identify shortcuts as a contributing factor
- Number of incidents during which safe operating limits were ignored or intentionally exceeded

- Number of incomplete shift logs or reports
- Number of missed surveillance rounds
- Number of work orders attributed to equipment abuse
- Number of times workers are challenged to solve "what-if" scenarios (assuming that this is part of the on-the-job training program)
- Number of persons who are past due on required reading
- Number of incidents involving disruptive personal behavior
- Number of disciplinary actions
- Percentage of workers failing random substance abuse tests

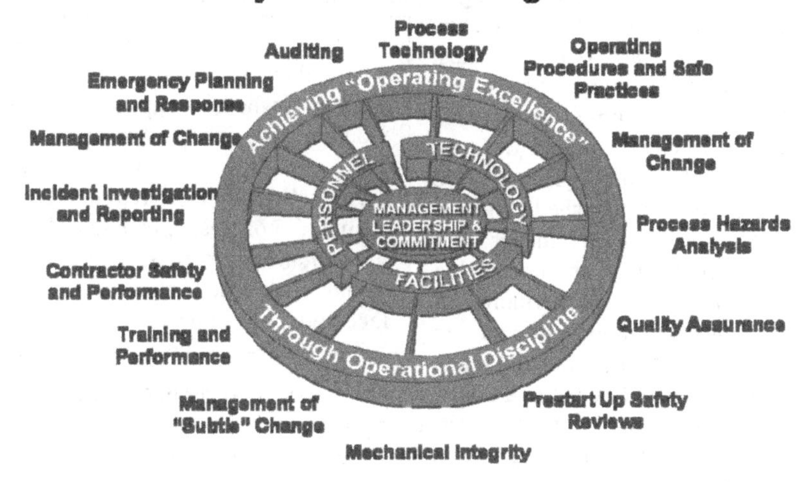

FIGURE 6.1. DuPont PSM Wheel

TABLE 6.1. Comparison of Activity-Based and Outcome-Based Metrics for Safe Work Practices

COO – Activity-Based Metrics	OD – Outcome-Based Metrics
Leading Indicators ▪ Compliance with the SWP authorizer training schedule ▪ Number of total hours spent in SWP training **Lagging Indicators** ▪ Number of incidents that included a failure of the SWP system as a causal factor	**Leading Indicators** ▪ Fraction of the SWPs that identified all applicable hazards and listed appropriate precautions/safeguards (as determined by the safety department during periodic worksite inspections) **Lagging Indicators** ▪ Number of unmitigated hazards or unsafe conditions identified during worksite inspections

These metrics provide a glimpse into the effectiveness of a facility's OD system, but they cannot be separated from the COO system. For example, incomplete shift logs or reports could be due to random errors made by individual operators (i.e., an OD failure), or they could indicate a lack of supervision or that management is growing more tolerant of deviations (i.e., COO failure). Thus, a change in metrics that track OD performance must be evaluated to determine whether the issue is systemic or related to individual behavior.

As described in Chapter 4, no program will eliminate human error. Human errors can be reduced by training workers thoroughly, managing worker stress and fatigue, providing logical human-machine interfaces, and so forth. In addition to these measures, an effective OD system should measurably reduce the likelihood that workers will accidentally or intentionally violate policies, procedures, and practices. An effective OD system will also produce an organization that is intolerant of intentional violations, regardless of the intent or outcome.

> Workers will confront situations where there simply are no rules, or situations where, if applied, the rules would clearly fail. OD, coupled with COO, equips members of the organization with the required tools, training, policies, procedures, *and understanding* necessary to make sound risk-based decisions when faced with uncertainty.

An effective COO/OD system demands that in all but the most extraordinary situations, members of the organization work within the system. Moreover, it provides workers with the knowledge and skills necessary to determine when the established system is failing, as well as a means for authorizing actions that are not within standard practice. Furthermore, it promotes making improvements to management systems rather than improvising fixes. The goal of the OD system is not to create human robots that perform tasks the same way every time, regardless of the outcome. Such an organization would not last long; it would be unable to cope with external and internal changes and be quickly overtaken by more creative competitors. Instead, an effective COO/OD system requires – some would say demands – that the organization continuously learn and improve, but that it does so in a disciplined manner. For example, improvements should be based on sound engineering principles, not merely on the fact that "It worked out okay the last time" or "It only shortens the prescribed heat-up cycle by 10%, and surely the designers allowed for that much variance."

The eight OD attributes described in this chapter are divided into two groups. Section 6.2 addresses four attributes that apply to the organization – what the organization expects of its leaders and the standards it sets for the work environment. Section 6.3 addresses four attributes that apply to individuals – attributes that shape behavior and help determine what workers do, or fail to do, on a daily basis. Sections 6.2 and 6.3 begin with brief introductions that describe their contents more fully.

The description of each OD attribute begins with a short summary, followed by an example that helps demonstrate the attribute. In some cases, the example is based on a high-profile incident. In other cases, the example is based on a combination of events that were never publicized outside of the facility where they occurred. The stories are not intended to summarize significant historical accidents. The CCPS and others have done that elsewhere (Refs. 6.3, 6.4, 6.5, 6.6, and 6.7). Rather, readers should use each story as an example to help determine whether the attribute is relevant to their operation; this can be done by asking questions such as "Could that happen here?" and "If it did happen here, would the consequences be tolerable?" Based on the answers to those and similar risk questions, readers are challenged to (1) further evaluate the lessons from the example, (2) determine whether a similar gap exists in their organization, and (3) evaluate benefits that might be provided by addressing the OD attribute in that section.

Before reading further, review Figure 6.2 on the next page, which is the COO/OD improvement and implementation cycle graphic that appears throughout this book. Many readers are interested in establishing a COO system (entering the cycle from the 12 o'clock position). The next step is to establish goals for each COO or OD element for key stakeholders to review, understand, and hopefully embrace. The goals need to be realistic and based on improvements that can be delivered by COO/OD. The goals should include some tangible benefit that will motivate the organization to spend time, allocate resources, or otherwise expend effort. To move forward, strong commitment and active management support will be required.

This chapter, along with Chapter 5, addresses the box in the 3 o'clock position. Sections 6.2 and 6.3 describe eight attributes, or ideas, that will help a facility establish and improve its OD system. The narrative description of each attribute can be used to benchmark existing programs or identify desirable features for new programs. Given the goals, the next step is to assess the current systems in place at a facility to identify which OD attributes are most likely to help the facility achieve the goals. This, along with ideas on how to implement a COO/OD system or elements of a COO/OD system, is addressed in Chapter 7.

6.2 ORGANIZATIONAL ATTRIBUTES

If leaders in some departments or work groups try to improve the OD in their area but there is no effort to implement an effective OD system elsewhere at the facility, the push toward higher standards will seem arbitrary and can cause significant organizational stress. Some workers will notice that "The guys in the adjacent unit are faced with similar hazards and risks, but they don't have to do all of this extra work," and will resist the changes that OD brings. Without leadership, management commitment, and transparency to set the stage for improvements in OD, the program will struggle and could even adversely impact the facility's culture and performance.

> It is hard for a work group to significantly outperform the rest of the organization on a consistent basis.

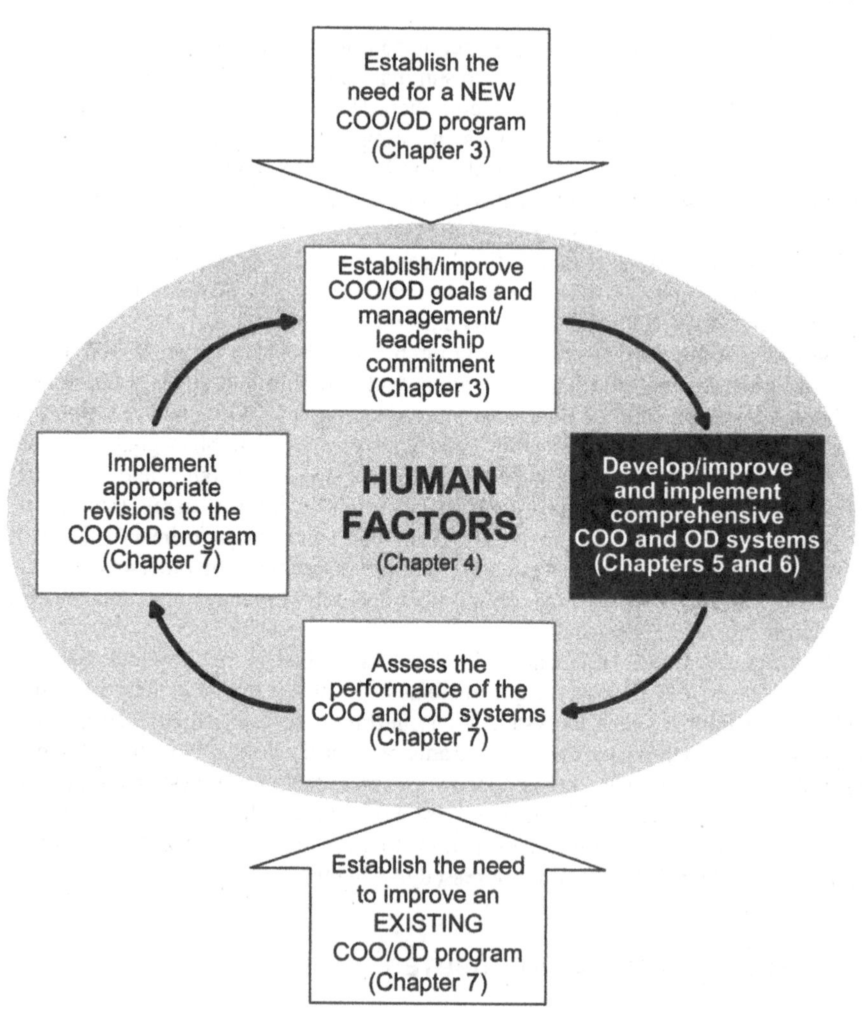

FIGURE 6.2. COO/OD Improvement and Implementation Cycle

Organizational attributes strongly influence the OD system, which in turn helps shape individual behavior. In fact, small organizations, where the leader knows and personally interacts with each individual, often develop an effective OD system simply through the leader's guidance and charisma. As the organization grows, the leader can no longer interact with everyone and, therefore, must depend on subordinates to help establish standards and enforce policy. In this environment, more formal systems are needed to help ensure consistent practice. At the simplest level, setting standards and consistently encouraging and enforcing them are COO and OD.

The organizational attributes that are critical to an effective OD system include:

1. Leadership
2. Team building and employee involvement
3. Compliance with procedures and standards
4. Housekeeping

6.2.1 Leadership

Leadership is the art of supporting and influencing others in pursuit of a common goal. As described in Chapter 5, COO includes numerous attributes for supporting the common goal of reliable performance. OD addresses specific behaviors that can be influenced by effective leadership.

> Early one morning, a maintenance worker who was assigned to replace the blades on a grinder slipped and cut his hand, requiring four stitches to close the wound. The procedure required leather work gloves, but he had removed them to start a nut on a stud bolt and had not put them back on before lifting the next blade into place. The incident report noted that "The employee knew he was supposed to be wearing gloves when he lifted the next blade; he simply failed to follow procedure." The worker didn't dispute the finding; in fact, he offered it up as the sole cause of the incident. The team quickly accepted his conclusion, and wrote two recommendations: (1) reinforce the glove policy and (2) explore better alternatives for hand protection while changing grinder blades.
>
> Shortly after lunch, the maintenance manager endorsed the incident investigation team's report and passed it along to the facility manager for approval. Later that day, the facility manager called the maintenance manager to the front office. Although the facility manager strongly supported the recommendation to explore alternatives for better hand protection, she rhetorically asked about other times that prescribed protective equipment was not being used properly, making the point that the incident investigation had not addressed why the organization continued to tolerate unsafe actions and unsafe conditions.
>
> The next day, the incident investigation team was reconvened, and it submitted a third, much more effective recommendation. The result of the final recommendation was that every leader at the plant, including the facility manager, committed to correcting any unsafe act or condition whenever it is discovered, and agreed that there is no meeting, phone call, or other activity that is so pressing that they would choose to not stop and take action to reduce the risk of harm to coworkers.

One important aspect of leadership – leading by example – is very closely associated with OD. Popular clichés such as "walk the talk" speak to our need to be led by those who demonstrate high regard for standards. The COO/OD system involves the entire organization, and it is hard to envision a case where it could be effective when leaders at any level or in any part of the organization fail to demonstrate the behaviors they demand from others.

Managers, superintendents, and supervisors are appointed by the organization and vested with power and authority. Leaders may be likewise appointed, but they receive their true power to lead from their work group. There are also multiple natural, unappointed leaders within most groups, and their support of the OD system will reinforce the group's values and significantly contribute to overall success. However, if unappointed leaders are not aligned with a new or modified OD system, this will create organizational stress and potentially undermine the OD system.

> A project manager decided not to adhere to a policy that required the spectacle blinds on all lines to be rotated prior to vessel entry. He felt his decision was justified because the entrants would be in the vessel a very short time. This was an OD failure on the part of the project manager and the entrants.

OD is most often associated with activities performed by operators and maintenance personnel and their direct supervisors. In most cases, supervisors will oversee day-to-day activities and choose whether to enforce standards or look the other way. However, everyone in the organization is routinely faced with the decision to enforce work rules, point out unsafe acts, correct unsafe conditions, and see that standards are enforced. Effective leaders choose to do all of these things.

OD goes well beyond decisions made by shop floor personnel and their immediate supervisors. Department heads, mid-level managers, and facility managers are routinely faced with hard decisions, where bending the rules might provide a short-term economic benefit or some other immediate payoff. The first inclination should always be to abide by established standards. However, rules cannot address every conceivable situation or condition, so it may be impossible or impractical to comply due to unforeseen circumstances. Thus, there must be a formal protocol to authorize one-time or permanent changes to the rules once due consideration is given to standards, risk, and the impact the change may have on the organization and the COO/OD system. When exceptions to established norms or rules are made, it is important to be transparent – to clearly communicate what is happening and describe how alternative safeguards are providing sufficient risk reduction until the special situation has been rectified.

OD applies to the work performed by many different groups. Leadership and enforcement of standards are not limited to operators, maintenance employees, and their supervisors. Although the standards may differ for support functions, they should be results-oriented and support the organization's goals and objectives. For example, very strict rules may govern the shift change time for operators because it is critical that full staffing be in place at all times. Similarly, it is critical that technical staff complete process safety action items as planned, and they should be held accountable for doing so. Just as an organization demands that operators adhere to standards for operating limits and limiting conditions for operation, it should demand that technical staff take a leadership role in managing risk. For example, technical staff should be held accountable for (1) allocating the capital funds and other resources needed to do the job right the first time and (2) conducting appropriate hazard identification studies and risk analyses prior to operating a process at plant scale, even though it might delay launch of a new product.

Effective leaders support the COO/OD system by:

- Setting the example
- Visibly demonstrating the value they put on the safety of everyone in the organization
- Providing sufficient and capable resources to do the job right
- Providing appropriate rewards and punishments (see the discussion of ABC analysis in Section 4.8.2)
- Being present in the work areas and actively engaging workers
- Understanding how decisions to circumvent previous rule-based decisions will be viewed, and proactively addressing the OD aspects of any such decision with stakeholders
- Applying OD widely and including the various functional groups that support operations
- Documenting, maintaining, and monitoring suitable metrics that characterize the OD system
- Recognizing and rewarding good OD performance

6.2.2 Team Building and Employee Involvement

Team building and employee involvement include a variety of management practices centered on empowerment and mutual trust that solicit workers' input on decisions and actions that affect their jobs. Employee involvement goes beyond employee participation. Employee participation may mean nothing more than including representatives on working, decision-making, or advisory teams such as PHA teams or the facility's process safety committee. Employee involvement activities take place continuously and engage all members of the organization.

In the early 1980s, the U.S. government sponsored a major research and development program to build an advanced gas centrifuge to significantly reduce the cost of enriching uranium. One potential bottleneck was the assembly of test centrifuges because that work required a very high degree of precision and skill. The assembly team, which soon became known to everyone as the "A team," clearly realized how the entire organization depended on its work. The team members took great pride in doing very high-quality work, in accordance with established procedures, and on the schedule required to support program objectives. Not only did the assembly team strictly follow procedures, each member contributed very effective ideas on how to improve centrifuge assembly. This significantly accelerated the development of methods for rotor assembly that had previously been left to the engineering/technical staff.

Employee involvement is not the goal; rather, it is the means to an end. It is based on the premise that workers are well qualified to help determine how their work can best be performed. Involvement facilitates continuous improvement and organizational success. Examples of employee involvement initiatives described in this section include:

- Day-to-day hazard identification and risk management programs such as job safety analysis (JSA)
- Peer-to-peer safety management programs such as DuPont's STOP™ (Safety Training Observation Program) and similar behavioral safety programs
- The Kaizen program
- The "5S process"

Job safety analysis has been used widely in the process industries for decades to identify the hazards of specific tasks and evaluate the controls in place to manage risk. An effective JSA program helps reduce risk. However, a JSA program that has degraded to the point where completing the paperwork is more important than identifying hazards and evaluating risk is poison to the OD system. It is also a clear sign that the facility is failing to maintain a sense of vulnerability, a cultural issue that lies at the root of many major process safety incidents.

Other day-to-day hazard identification and risk management activities include **peer-to-peer safety management programs** and more formal **behavioral safety programs**. Both approaches have a high degree of employee involvement and are proven to be very effective. However, they require strong leadership and management support to remain effective. Otherwise, they too may become "one more thing we know we are supposed to do but don't do very well anymore." From an OD perspective, this is another step on the path to failure.

Kaizen is a system of continuous improvement that can be used to address quality, technology, process, productivity, and safety issues; it has even been used to address culture and leadership concerns (Ref. 6.8). The word "Kaizen" roughly translates from Japanese as "good change." Kaizen strives for incremental improvements on a regular basis. While some facilities believe that "If it isn't broken, don't mess with it," the Kaizen philosophy is to "Improve it even if it isn't broken, because if we don't, we won't be able to compete with those who do." The Kaizen cycle includes the following steps:

1. Standardize an operation
2. Measure the standardized operation
3. Gauge measurements against requirements
4. Innovate to meet requirements and increase productivity
5. Standardize the new, improved operations
6. Repeat the cycle

The process industries have been using a form of Kaizen for several decades. PHA teams routinely examine processes to evaluate risk associated with deviations from design intent. Based on their analysis, teams often make recommendations that provide incremental risk reduction, even if the postulated scenario has never occurred at the facility. In addition, PHAs are cyclical events; they are periodically reviewed and updated. Standardizing operations, evaluating how they run, identifying opportunities to improve, implementing approved changes, and incorporating them into the "new normal" are all part of the life cycle of an effective process safety program. This same approach can help facilities identify and effectively implement improvement ideas that support a variety of objectives.

Note the similarities between the Kaizen process, PSM systems, and COO/OD. The Kaizen process includes steps for standardizing and measuring (metrics); understanding and comparing the status quo to requirements (understand risk significance); innovation, not by individual freelancers but by a well-controlled change and improvement process (management of change); and standardization of the new/improved process.

The **5S process** is designed to create and sustain a safe, productive workplace. As with all of the other formal processes described in this section, it is implemented by teams of involved employees.

One reliable measure of teamwork and employee involvement is worker pride. Contributing to the success of the organization in a meaningful way motivates employees and increases the pride they have in their job, work group, organization, and employer. Proud workers also project a positive image of their employer and workplace to others in the community.

All of these processes, as well as any other continuous improvement processes, have several features in common:

- Improvements come from teams of involved employees. Organizations may use outside facilitators to "get the ball rolling," but ownership for the employee involvement process lies within the line organization.
- Each employee involvement process requires management effort to keep it going, particularly until it becomes part of the organization's culture or "the way we do things around here." In that respect, any activity that is not essential to delivering product will suffer when management stops paying attention. The resulting difference between what is described on motivational posters in the break room and what really happens in the facility is a threat to the OD system.
- Improvements almost always come in many small steps. Sustained effort rather than "finding the silver bullet" most often yields the greatest improvement.
- Workers who are most at risk are actively, or better yet, passionately involved in efforts to improve safety.
- Workers are proud of the contribution they make to the organization.
- Workers' contributions to improving the organization are widely recognized and appropriately rewarded.

6.2.3 Compliance with Procedures and Standards

Section 5.5.8 addresses the COO attribute of intolerance of deviations from an organizational standpoint. This section addresses compliance from an individual or small work group perspective. Few in an organization may know about a certain shortcut, but at least one person does. In some cases, supervisors and others "don't want to know" how the job got done, choosing to congratulate the work crew rather than confront an unauthorized, and possibly unsafe, action.

Section 4.5 describes the difference between rule-based, knowledge-based, and skill-based work activities and notes that the same worker is often expected to work in all three domains in a single procedure. An effective COO system recognizes when activities should be rule-, knowledge-, or skill-based, and provides appropriate knowledge, skills, and resources. This section focuses on compliance, which is most closely related to rule-based policies and procedures.

> Commercial pilots use rule-based procedures for pushing off from the gate, taxiing to the runway, and taking off (for example, there are prescribed settings for the flaps and there is a minimum speed for rotation). However, both skill and knowledge are required to successfully move through the checkout, push-back, taxi, and takeoff sequence at different airports under different weather and traffic conditions.

One class of deviations is the shortcut. Workers often discover alternate ways to get a job done. Some shortcuts make the job easier; others improve productivity. In either case, unauthorized shortcuts should not be tolerated, even if there are short-term benefits. At some facilities, workers who come up with faster/better methods receive positive recognition for their ability to "get the job done." In the absence of OD, management personnel intention-ally turn a blind eye toward what workers do because they are only interested in achieving the desired results.

> A facility's policy allowed railcar unloaders to go home when all cars had been unloaded for the day. On occasion, high level in the storage tank tripped an interlock that stopped the unloading process. If the unloaders were confident that there was room in the tank, and they did not want to wait for the tank level to fall as the material was used in the process, they overrode the interlock. Eventually, this resulted in an overfill. The overfill was partially caused by a policy that effectively rewarded workers for taking shortcuts.

A second class of deviations is the error of omission – a skip is stepped or something doesn't get done. It may be that workers discover that if they do nothing, the outcome is normally just fine. For example: (1) operators may find that they can skip a set of readings each shift with no adverse outcome or (2) the lubrication mechanic may discover that "hard-to-reach" lube points only need a shot of grease every other week. Other safeguards take up the slack. Besides, it is unlikely that whatever event is required to initiate a loss event will occur before the end of the shift. There is probably no downside to not following the rules. Just as the vast majority of people who choose to not wear seat belts arrive at their destination unhurt, the vast majority of workers who fail to perform simple maintenance checks, operator rounds, and similar routines will complete their shift without incident. OD demands that the tasks be performed regardless of the worker's belief that everything will be okay even if the tasks are not performed.

A third class of deviations stems from failure to provide sufficient time, adequate tools, or other resources. Although adverse outcomes due to these factors are too often blamed on individual employees as an OD failure, they really stem from an inadequate COO system. This topic is addressed in Section 5.4.4.

A more subtle class of deviations is failure to hold coworkers accountable for following established policies and procedures. Workers can be reluctant to confront or report a coworker. However, in the process industries, where a single error can initiate a chain of events that leads to multiple serious injuries or fatalities, it is critical that workers look out for one another. Tolerating shortcuts or other errors by coworkers can put everyone's safety in jeopardy.

When addressing shortcuts and errors of omission, it is important to dissociate the outcome from the activity. If an unauthorized deviation to a rule-based policy or procedure resulted in a positive outcome, such as avoiding a unit shutdown or returning the unit to normal operation more quickly, the consequence for the worker who chose to deviate from the policy/procedure should still be based on what actions were taken, not the positive outcome. Likewise, if the result had turned out very badly, such as a runaway reaction, the consequence for the worker who chose not to shut down when prescribed limits were exceeded should be no more or no less severe than the last time the same action was taken by another worker when there was no adverse outcome.

> Dissociating activity and outcome can be very difficult. For example, few would argue that a worker should hold his breath and enter a confined space under any circumstances, even to rescue a coworker. Although it would be difficult for most people to describe the rescuer who risked his life to save another as anything other than a hero, confined-space incidents often do result in two fatalities, and the rescuer's actions in this hypothetical case are imprudent. It is essential that workers have the discipline to adhere to procedures, including confined space entry procedures, in spite of the strong instinct at times to do otherwise.

Yet another class of deviations stems from the belief that "it's always been that way," a classic symptom of normalization of deviance. Unsafe acts performed on a routine basis go unnoticed or unchecked. For example, on September 14, 1997, a worker at the Hindustan Petroleum Corporation Ltd. (HPCL) refinery opened the drain valve on an LPG sphere and went on to perform other tasks, as it normally took several minutes to drain the water from the sphere. The operator was not aware that the sphere had been recently drained. The resulting release of LPG set off a chain of fires and explosions that resulted in sixty fatalities. Tolerance of repetitive deviations undermines the OD system (Ref. 6.9).

The initiating event for the 1997 HPCL refinery disaster and many other disasters is human error. A more complete discussion of human error is included in

Chapter 4. Simply put, an OD system, or any other initiative, will not eliminate human error. A deviation is not indicative of an inadequate OD system, but tolerance of an unacceptably high error rate does point to OD, culture, and possibly COO concerns.

Activities that help identify and eliminate deviations include:

- Planned job observations for critical, routine tasks. Supervisors tend to closely monitor unusual events but not check on routines until something goes wrong. Planned job observations can help identify shortcuts, eliminating ones that are deemed unsafe and potentially improving policies and procedures for those that are eventually determined to be an improvement.
- Random checks for routine maintenance and similar activities. Some violations are easy to spot, such as someone not operating equipment properly. Others require a closer look. For example, if the operator is required to evaluate the condition of the equipment using a prescribed checklist prior to each use, an effective OD system would have supervisors reviewing the checklist at random intervals to ensure that it is completed properly and completely, and at the time the activity is performed.
- A minimum number of random checks per month of compliance with work permits. These checks are often assigned to operators, maintenance workers, and members of the technical staff to better distribute the workload and avoid the "there's the safety guy, better straighten up" reflex.
- Refresher training programs that reinforce the need to perform activities as described, and the potential consequences of failing to do so.
- Continued emphasis on the basis for safety, including elimination of unsafe conditions.

6.2.4 Housekeeping

Housekeeping is an indicator of the effectiveness of the OD system. However, like employee involvement, it is a means to an end rather than the end in itself. It is one of many factors that contribute to a safe and productive work environment. A cluttered work environment can lead to industrial safety accidents, and it almost certainly reduces productivity. However, there are process safety aspects as well. For example, if equipment, floors, and other surfaces are kept clean, it is much easier to identify small leaks before they become a major release.

OD goes well beyond housekeeping. If the facility's standard is to cap the end of drain and bleed lines, caps are in place. Likewise, machine guards are installed,

> A security guard on routine rounds noticed a leak. He knew it was a problem because nothing ever leaked; any puddle on the floor was a problem. He called the control room. The call resulted in saving the plant from running the pump to failure.

pipe hangers are in place, there are no open electrical cabinets or uncovered electrical boxes, fire doors are kept closed, and so forth. Anything not properly closed or sealed is obvious to anyone who surveys the area.

The most immediate indicators of good housekeeping are (1) lack of clutter; (2) clean walls, ceilings, and floors; and (3) a sense that things are put where they belong. When deviations are found, organizations that embrace OD fix the unsafe condition and proactively look for other locations where the same problem might exist. Even more importantly, they ask themselves how the unsafe or unproductive condition was allowed to develop and persist, and they take action to address the OD issues that are identified.

6.3 INDIVIDUAL ATTRIBUTES

This section addresses several attributes that are internal to the individual members of the organization. As a senior facility manager once said, "You live and die with your people." Excellent programs, policies, procedures, etc., do not ensure excellent operation. Excellence depends on how well people do their jobs.

The attributes described in this section are inherent in the person who is hired for a job. However, they can be contagious – placing value on them causes everyone to try to do better or enhance their abilities. The individual attributes addressed in this section include:

1. Knowledge
2. Commitment
3. Awareness
4. Attention to detail

6.3.1 Knowledge

Training and development of competence are vital parts of a COO system (see Section 5.5.4). Workers need to understand how to do a task. However, they also need to understand certain aspects of how things work and why performing the task as written is important. If the process or equipment they work with is a big black box, they may fail to recognize when things are no longer "within the lines" that define the boundary for safe, proper operation. Finally, workers need to recognize when they have reached the limit of their knowledge and when to seek assistance from supervision or the technical staff. In facilities that have an effective COO/OD system, such requests are viewed positively, not as an admission of inadequacy.

Training needs should be segmented among operators (users), routine maintainers, repairers, and designers. For example, the automated teller machine (ATM) revolutionized how we withdraw cash from bank accounts. Users had to learn steps that were simple and relatively intuitive, but completely different. If the steps were not followed exactly, the machine did not dispense cash. New jobs with different training requirements were created. Someone had to restock cash in the ATM, and this required some knowledge of how the ATM worked. The "master mechanic" for the ATM machine – the person who was called in when the problem was not routine – needed to know enough about the machine's design to properly diagnose and fix any problems. Of course, the chief designer and the design team needed to know quite a bit about machine design, mechanics, properties of materials of construction, optical imaging, and so forth. Each individual or work group had a different knowledge requirement, and training programs needed to be developed to address these specific needs.

The same model applies throughout the process industries. However, the consequence of failure can be much more severe than not dispensing cash; it can be catastrophic. If a worker's understanding is limited to a memorized set of rules, he may not be able to recognize conditions where the rules do not apply, might fail to work, or even turn a process upset into a disaster. Knowledge better equips workers to spot deviations and to quickly and accurately assess the significance of the deviation.

During a 2009 interview with CBS News, Captain Chesley Sullenberger insisted that landing U.S. Air flight 1549 in the Hudson River was a matter of "following procedures." According to Captain Sullenberger, the first officer did his part by trying to restart the engines and adjusting flap positions while Captain Sullenberger decided where to land and continued to fly the plane. In other words, according to Captain Sullenberger, ditching in the Hudson was simply complying with a set of established procedures.

When interviewed shortly after the landing, Captain Sullenberger said, "One way of looking at this might be that for 42 years, I've been making small, regular deposits in this bank of experience: education and training. And on January 15 [2009] the balance was sufficient so that I could make a very large withdrawal." (Ref. 6.10)

Effective hiring and training programs help ensure that workers have a common set of required knowledge, and then allow them to take advantage of other opportunities to expand their knowledge, if desired. People like Captain Sullenberger, who have an inherent desire to expand their process knowledge, are almost always more effective employees.

Maintaining organizational competence, training effectively, following policies and procedures, and reporting unusual conditions are all part of an effective COO/OD system. However, for the system to succeed, there needs to be (1) someone who possesses the knowledge and (2) someone who decides that they need this knowledge. The second factor, the "demand side" of the equation, is not always present. We cannot rely solely on individuals to identify these needs. Moreover, since many processes are operated by a team, it is helpful for all of the team members to have a common understanding of the process. Thus, facilities need to intentionally identify and transfer critical knowledge to workers. Effective leaders do not just facilitate knowledge transfer; they actively encourage workers to advance their skills and knowledge, and they model this behavior themselves throughout their career.

There are many ways to transfer knowledge. Some, such as instructor-led courses, are relatively effective but costly. Others, such as asking people to review procedures, unit technology manuals, or similar documents in their spare time, are very inexpensive but relatively ineffective. (One author in the field of COO/OD reports that educational studies have shown that we only retain 10% of the information we read [Ref. 6.11].) Generally, a higher level of engagement leads to better learning. One of the benefits of having operators participate on hazard analysis teams is being exposed to the learning and retention that take place in an environment that is much more interactive than conventional training. Rotating participation will at least train workers to think in "failure space" (i.e., "What would happen if . . .") rather than "success space" (where they act and then the intended consequence occurs).

Hands-on training can enhance understanding, but it normally requires more time and effort. Consider fire-extinguisher training. One training method is to have everyone read a short summary of the conditions necessary for a fire (fuel, oxidant, ignition source, and mixing) and how to operate a fire extinguisher. A second is to hold classroom training, possibly incorporating a demonstration where the instructor simulates putting out a fire. A much more reliable method is to supplement the classroom training with an exercise in which trainees extinguish a real fire under safe/controlled conditions. Many process facilities send their emergency responders to special schools specifically to gain this sort of real-world experience. In this way, rule-based information is reinforced by practice.

Effective knowledge transfer needs to be planned and scheduled. In other words, organizations cannot wait for someone to wake up one morning and decide that he or she does not have a sufficient understanding of fire or is not proficient in operating fire extinguishers. People rarely do that. And, certainly, organizations cannot wait until there is a fire to have someone determine that "It would be really handy to be fully trained on the proper use of that fire extinguisher."

In some cases, a significant amount of the knowledge-based training is ad hoc. Workers learn from (1) their own experience, (2) incidents they happen to witness, (3) stories and recollections about past events, and (4) conversations with more senior colleagues or personnel from other departments or facilities. Experience is the best teacher, but not necessarily the best way to learn. There is no assurance that all employees obtain the required knowledge through ad hoc on-the-job training. In addition, they may gain incorrect or out-of-date knowledge. And mistakes (the most memorable kind of experience) may be costly.

Successful organizations closely integrate rule-based and skill-based training with knowledge-based training. The former address the question "What are we supposed to do and how?" Knowledge-based training addresses the question "Why do we do it this way and why is that likely to work better than alternatives I might come up with?" Knowledge-based training provides a sound basis for (1) selecting between alternate actions to address a particular situation, (2) evaluating whether the selected action is achieving the intended result, and even (3) expanding rule-based procedures and skill-based training to address situations that are not explicitly addressed via procedures and training.

Key considerations when developing programs for improving knowledge include the following:

- Required reading is quick and inexpensive, but generally not effective.
- Instructor-led courses provide an opportunity to address questions and to observe body language to gauge participants' understanding; however, it is most effective (1) for knowledge-based training and (2) in cases where the instructor has sufficient knowledge and subject matter expertise.
- One-on-one discussions are often a good way to impart knowledge, but they tend to be random compared to structured courses. Also, there is a higher likelihood that incorrect or incomplete information will be provided.
- Integration of knowledge-based training with skill-based training, or activities that reinforce the concepts presented in knowledge-based training, improve the effectiveness of the knowledge transfer.
- Employees should question whether they have the knowledge to do the job correctly and safely, and be encouraged to highlight their individual training needs.

6.3.2 Commitment

What is the proper way to measure commitment? Is it the willingness to place yourself in harm's way for the greater good of the company? That is certainly not the intent. Is it commitment to a work group or a willingness to cover for other workers? Again, that is not the type of commitment that supports the OD system.

Within the framework of OD, workers personally commit to performing duties in accordance with prescribed policies, procedures, and practices. They also commit to working within the system to continuously improve policies, procedures, and practices. Workers commit to trusting colleagues and to clearly understanding concerns and issues rather than simply questioning motives. Most of all, managers commit to holding workers accountable for their actions, rather than the consequences of their actions.

Personal commitment does not imply unquestioning obedience. Nor is it measured in terms of good intentions. Instead, it is a set of actions (or a decision not to act) that are significantly influenced by the organization's fundamental principles. In the example that follows, if worker safety is fundamental, the day-shift maintenance crew members were more personally committed to the organization's principles. If, on the other hand, production topped safety (which was certainly not the policy or intention of this specific company or the process industries in general), the supervisors and second-shift workers were the more committed ones.

OD demands that management be likewise personally committed to address the fundamental attributes of OD. Careful attention should be paid to what is recognized and rewarded and what might result in adverse consequences for those involved. This is a good indicator of what management and the entire organization are committed to and what is valued.

A dip tube that was manufactured from a pipe welded to a collar designed to fit between two flanges was installed in a vessel in hydrofluoric acid (HF) service. The end of the pipe had corroded and failed. Although the facility's policy required that the HF tank be drained and cleaned out prior to being opened, the operations group determined that draining the tank would result in an unacceptably long outage and decided it would be safe to open the tank for a limited amount of time because the vacuum system would prevent HF fumes from exiting the open tank.

Two day-shift maintenance workers were assigned to remove what was left of the existing dip tube and insert the replacement tube. They balked, citing concerns about violation of the long-standing policy and a general concern about using an active safeguard such as the vacuum system to protect them while working atop a tank containing HF. The replacement dip tube sat outside, warming in the sun all day, while equally heated discussions continued between the maintenance workers and their supervisors. Sometime after 4:00 p.m., a much less experienced team of workers from the afternoon shift maintenance group agreed to insert the dip tube. All went as planned until the tube broke the surface of the liquid, simultaneously isolating the vacuum system and heating the acid enclosed in the open pipe. HF quickly boiled, badly burning one of the maintenance workers.

Clearly, hazards were not recognized by the supervisors and the afternoon shift maintenance workers. Even the day-shift maintenance crew members were unable to articulate the basis for their specific concerns; they were simply uncomfortable with working with a tank that contained HF, a previously forbidden activity. But it is also fair to examine the commitment of the three groups: (1) the day-shift maintenance crew who were committed to following policy and had a vague, but real, concern about doing the job they were assigned, (2) the supervisors who were committed to getting the plant back in production as quickly as possible regardless of plant policy, and (3) the afternoon crew who were equally committed to getting the dip tube inserted so that the plant could be restarted. Plans, as well as the justifications put forth to support the plans, often reveal much about commitment.

Another aspect of commitment is personal accountability. High-performing work teams hold each other accountable. Less effective work groups typically accept sloppiness and routine deviations and believe that the resulting adverse outcomes are simply bad luck. Individuals, work groups, departments, and entire organizations should be held accountable for what they can control and should, likewise, assume significant accountability for things they can influence.

Commitment is not easy to measure. It is hard to know how strongly an individual or group will commit to a task or goal. However, people's past performance is normally a good predictor of their future performance. That adage is particularly true of commitment – strong commitment leads to desired outcomes, which in turn increases individual and group commitment.

6.3.3 Awareness

Awareness is a key attribute for all members of an organization, from the CEO to front-line personnel. Awareness involves (1) perceiving the cues in your environment, (2) interpreting the meaning of those cues, and (3) projecting what will happen within the system in the future based on your interpretation. Thus, awareness is very important; unnoticed or unaddressed minor problems can lead to process upsets and eventually incidents in the absence of worker awareness.

A senior maintenance technician was assigned the task of replacing a level sensor in a formaldehyde tank. The tank had already been drained. Because they were short on operators and he was the most senior person in the plant, the technician convinced the operator to let him just go out and get the job done on his own. There were two identical tanks. The technician approached the tanks from the opposite direction he normally would. He went to the wrong tank and removed the face-high level sensor. The tank was almost full, so formaldehyde poured out of the tank onto his face. He stumbled and flipped backwards over the containment dike. He was found there by a staff person who happened to be walking by.

In some cases, individuals actively avoid awareness because it forces tough choices. Senior management can surround themselves with others of like mind, erecting walls that isolate them and lead to very poor decision-making. Facility management can withdraw to the front office, trying to avoid leaving their "fingerprints" on a failed policy or bad decision. On the front line, awareness often determines whether a process upset results in a minor hiccup, a brief outage, or a major incident. Awareness includes self-checking and peer-checking. The criterion is not whether it is "good enough," but whether it "meets the standard."

When faced with multiple, differing readings, instead of accepting the reading that is within the specified range and ignoring the others, aware workers determine which reading is correct, take action to stabilize or adjust the process as required, and then investigate and address (or initiate action to address) the cause of the erroneous reading. Moreover, aware workers understand hazards and risk, and they examine more closely anomalies that appear to be more risk significant.

Workers cannot take action to address an abnormal condition without first being aware of the situation. When faced with a highly structured task, people often have a tendency to tune out all input that does not appear to be related to the task at hand. Personnel who are keenly aware of their surroundings quickly spot unusual noises, odors, vibrations, or other patterns. Sometimes these are of little interest, but they often point to underlying problems that, in time, could become significant.

Situational awareness includes concerns involving adjacent units and work areas. Here are some examples:

- Catch tanks and vent stacks presented significant hazards to assigned work groups that were in no way connected with the unit where an incident occurred.
- A group of painters were startled when operators vented steam into the normally unoccupied area where the painters were working.

- A high concentration of water in an intermediate product led to a significant loss because the material was being diverted to a process that did not normally receive/consume the material.

Awareness should be emphasized any time there has been an interruption or turnover from one group to the next. In particular, if there are interruptions in nonroutine activities, workers should allow extra time to thoroughly familiarize themselves with the current status of the system. When handoffs are required or designed into the operation (such as shift turnover), it is important to have a structured procedure to heighten awareness, typically supported by checklists or written logs.

Finally, awareness alone is not sufficient. Noticing that a coworker is working at height without fall protection does not reduce risk. One must take action based on what has been observed. Everyone, from the facility manager to a temporary cafeteria worker, should understand their duty to report and/or take action to remedy any unsafe or unexpected situation.

Some workers are better able to sense conditions that are outside the norm. Whether they can hear frequencies that most others cannot, have a keener sense of smell, or are simply more observant, they are more likely to notice anomalies. Others become so focused on a task that they notice little else. These are truly personal attributes that can be more difficult to learn than some of the others discussed in this chapter. However, (1) simply making personnel aware of the hazards that surround them, (2) continuously reinforcing the need to be vigilant and aware, and (3) celebrating the loss event that never happened due to the actions of an alert employee help promote awareness.

6.3.4 Attention to Detail

Virtually everyone has been told that "If something is worth doing, it's worth doing right" or "The devil is in the details." Reliable performance requires attention to detail.

Attention to detail complements awareness. While a high degree of awareness might prompt workers to look around to see what else is happening that might affect their work, attention to detail requires that workers focus on the task at hand. These two attributes are not in conflict; rather, they are complementary.

OD demands precise and repeatable work, particularly if the consequence of making an error could be significant. COO complements this need by helping to identify critical tasks and practical measures that can prevent a single human error from leading to unacceptable consequences.

On September 8, 2004, NASA's spacecraft Genesis crashed into the Utah desert after spending three years in space collecting samples to support NASA's solar wind research program. The intent was for the capsule to enter the Earth's atmosphere, deploy a drogue parachute, and be "caught" in the air by a helicopter to protect the valuable samples. Instead, the parachute never deployed and the capsule hit the ground at 193 mph. It was later determined that a G-switch, designed to initiate deployment of the parachute based on the rate of deceleration, failed because the design called for it to be installed upside down. Presumably, due care was taken to ensure that the installation was correct, but insufficient care was taken to check the design. (Ref. 6.12)

Similar to any other risk management activity, safeguards that help prevent or mitigate the consequences of human error will fail some fraction of the time. Thus, reducing the frequency with which these safeguards are challenged reduces risk, and efforts to reduce errors should be considered alongside efforts to add more protective layers. (In many cases, a combination of reducing the error rate and improving safeguards is the optimal solution.) An effective OD system promotes efforts to reduce human error by avoiding the temptation to dismiss a significant error as a single human failure. Rather, the causes of error, including inattention to detail, are evaluated and steps are taken to retrain or otherwise reinforce the need to reduce error rates.

COO identifies critical tasks; OD is the precise and repeatable execution of those tasks. Attention to detail is apparent in every aspect of OD. It is filling every blank on a form correctly. It is mopping up spilled coffee in the hallway. It is ensuring that every word is spelled correctly. It is arriving at work on time. Some individuals are much more attentive to detail than others. Not everyone has the temperament to be an accountant or a surgeon. Thus, the COO system must highlight areas where attention to detail is critical to safety so that OD can ensure that everyone is attentive to those details.

6.4 SUMMARY

This chapter reviewed several organizational and individual attributes that form the foundation of an effective OD system. Given the diversity of facilities, some attributes will be more significant than others and, in a particular situation, some may not matter much at all. However, readers should review each attribute and determine (1) whether it is likely to result in a significant improvement in operations at their facility and, if so, (2) whether additional foundational work is required to support OD.

Knowledge of the attributes of an effective COO system, which were presented in Chapter 5, and of the organizational and individual attributes that support OD, which have been presented in this chapter, should help readers decide how COO/OD can benefit their organization. However, real benefits in safe and productive operations only come from embracing this approach and changing how things are done at a facility. The next chapter examines how the concepts presented in this book can be implemented effectively.

6.5 REFERENCES

6.1 U.S. National Transportation Safety Board, *Collision of Metrolink Train 11 with Union Pacific Train LOF65-12, Chatsworth, California, September 12, 2008*, Accident Report NTSB/RAR-10/01, PB2010-916301, Washington, D.C., adopted January 21, 2010.

6.2 Klein, James A., and Bruce K. Vaughen, "A Revised Program for Operational Discipline," *Process Safety Progress*, American Institute of Chemical Engineers, New York, New York, Vol. 27, Issue 1, March 2008, pp. 58-65.

6.3 Kletz, Trevor, *What Went Wrong? Case Histories of Process Plant Disasters, Fourth Edition*, Elsevier, Burlington, Massachusetts, 1999.

6.4 Atherton, John, and Frederic Gil, *Incidents That Define Process Safety*, Center for Chemical Process Safety of the American Institute of Chemical Engineers, John Wiley & Sons, Inc., Hoboken, New Jersey, 2008.

6.5 Kletz, Trevor, *Lessons from Disaster: How Organizations Have No Memory and Accidents Recur*, Gulf Publishing Company, Houston, Texas, 1993.

6.6 Kletz, Trevor, *Still Going Wrong! Case Histories of Process Plant Disasters and How They Could Have Been Avoided*, Butterworth-Heinemann, Burlington, Massachusetts, 2003.

6.7 Lees, Frank P., *Loss Prevention in the Process Industries: Hazard Identification, Assessment and Control, Second Edition*, Butterworth-Heinemann, Oxford, England, 1996.

6.8 Imai, Masaaki, *Kaizen: The Key to Japan's Competitive Success*, McGraw-Hill/Irwin, New York, New York, 1986.

6.9 Khan, Faisal I., and S. A. Abbasi, "The World's Worst Industrial Accident of the 1990s, What Happened and What Might Have Been: A Quantitative Study," *Process Safety Progress*, American Institute of Chemical Engineers, New York, New York, Vol. 18, Issue 3, Autumn 1999, pp. 135-145.

6.10 Interview transcript, *60 Minutes*, CBS News, February 15, 2009.

6.11 Howlett, H. C., II, *The Industrial Operator's Handbook: A Systematic Approach to Industrial Operations, Second Edition*, Techstar, Pocatello, Idaho, 2001.

6.12 National Aeronautics and Space Administration, *Genesis Mishap Investigation Board Report, Volume I*, Washington, D.C., October 2005.

6.6 ADDITIONAL READING

- Buckingham, Marcus, and Curt Coffman, *First, Break All the Rules: What the World's Greatest Managers Do Differently*, Simon & Schuster, New York, New York, 1999.
- Byham, William C., Ph.D., and Jeff Cox, *Zapp! The Lightning of Empowerment: How to Improve Productivity, Quality, and Employee Satisfaction*, The Ballantine Publishing Group, New York, New York, 1998.
- U.S. Department of Energy, DOE Order 5480.19, Change 2, *Conduct of Operations Requirements for DOE Facilities*, Washington, D.C., October 23, 2001.
- Goman, Carol Kinsey, Ph.D., *Managing for Commitment: Building Loyalty Within Organizations*, Crisp Publications, Inc., Seattle, Washington, 1995.
- Klein, James A., "Operational Discipline in the Workplace," *Process Safety Progress*, American Institute of Chemical Engineers, New York, New York, Vol. 24, Issue 4, December 2005, pp. 228-235.
- Senge, Peter M., *The Fifth Discipline: The Art & Practice of the Learning Organization*, Doubleday, New York, New York, 2006.

7
IMPLEMENTING AND MAINTAINING EFFECTIVE COO/OD SYSTEMS

7.1 INTRODUCTION

Chapter 3 described the upper-management leadership necessary to establish COO/OD goals and initiate and sustain the management system. This chapter describes the role of managers and supervisors in developing, implementing, and maintaining an effective system. It relies heavily on Plan-Do-Check-Adjust (PDCA), which is a cyclical, four-step approach for implementing business process changes such as COO/OD. Sustained management effort is required to drive that

NUMMI – An Example of Implementing Effective Programs

In 1983, the Toyota Motor Corporation (Toyota) and the General Motors Company (GM) began a joint venture, New United Motor Manufacturing, Inc. (NUMMI). Toyota wanted to rapidly begin building cars in the U.S., and GM wanted to learn more about Toyota's production system and restart the idle Fremont, California, plant.

However, the Fremont plant had been closed for a reason. Its workforce had a horrible reputation for going on strike frequently, filing repeated grievances, and producing poor-quality products. Absenteeism often exceeded 20%. Toyota had many concerns about whether workers with such a bad reputation could embrace the teamwork and employee participation concepts that are central tenets of its production system.

Nevertheless, Toyota confronted the Fremont realities, planned a training program for the American employees of NUMMI, and proceeded to teach them its production system. It cultivated employee involvement and strove to develop a climate of mutual trust between the workers and management, even going so far as to involve the production floor leaders in the hiring of their team members. As problems arose, they were resolved with employee input. When sales were slack in the late 1980s, NUMMI cut plant operating hours and management bonuses to avoid worker layoffs – reinforcing management's commitment to the workers and their welfare.

The results were astonishing. Within a year, absenteeism plummeted to about 2% while product quality skyrocketed from the worst in GM to the best. The workers, the union, and the plant were the same. By changing the management and production systems and implementing an effective COO/OD system, NUMMI achieved its organizational objectives for all its stakeholders. (Ref. 7.1) process, both initially and through subsequent iterations of the PDCA cycle for continuous improvement. Management must also ensure that the resources it commits can realistically achieve the goals it sets.

During the **Plan** phase, the objectives and processes necessary to deliver the expected output are established. Thus, a complete and accurate specification of the expected COO/OD system outputs is essential to developing a successful plan. The Plan phase also includes the selection of metrics that will be used in later phases.

During the **Do** phase of the cycle, the processes are implemented. Small-scale pilot testing is usually beneficial even though it will require multiple loops through the PDCA cycle. Nevertheless, it generally results in a more efficient (and less disruptive) implementation of the COO/OD system.

During the **Check** phase, the processes are assessed against specific goals using selected metrics and investigation results. The results are compared to baseline values to ascertain any differences.

During the **Adjust** phase, the measured results are compared to the expected results, and any differences are analyzed to determine their cause and significance. During this phase, management decides where to apply further changes to improve the COO/OD results. Even if the results are satisfactory, there may be opportunities to improve the efficiency of the process.

Figure 7.1 illustrates the PDCA cycle as applied to COO/OD. Organizations implementing a new COO/OD system, as discussed in Chapter 3, would enter the PDCA loop at the top of the diagram during the **Plan** phase; those refreshing or improving an existing system would enter from the bottom during the **Adjust** phase. When a cycle through the first three phases does not identify a need for adjustment or an opportunity for improvement, the objectives should be refined to promote continuous improvement. The four phases should then be repeated periodically.

Table 7.1 lists the basic steps in the PDCA process when it is applied to COO/OD implementation. These steps will be discussed further in this chapter.

7.2 DEVELOP A PLAN

As discussed in Section 3.3, upper management's first task is to establish expectations. What are the ultimate goals? What is upper management's vision of success? Chapters 5 and 6 describe the building blocks for a COO/OD system.

When management is developing its plan and deciding what to attack, those two chapters provide ideas on where to start. Once the goals are understood, managers must then develop SMART[8] action plans that they reasonably believe will achieve the organization's goals, considering the reality of the current situation.

[8] SMART – Specific, Measurable, Attainable, Relevant, Time-specific

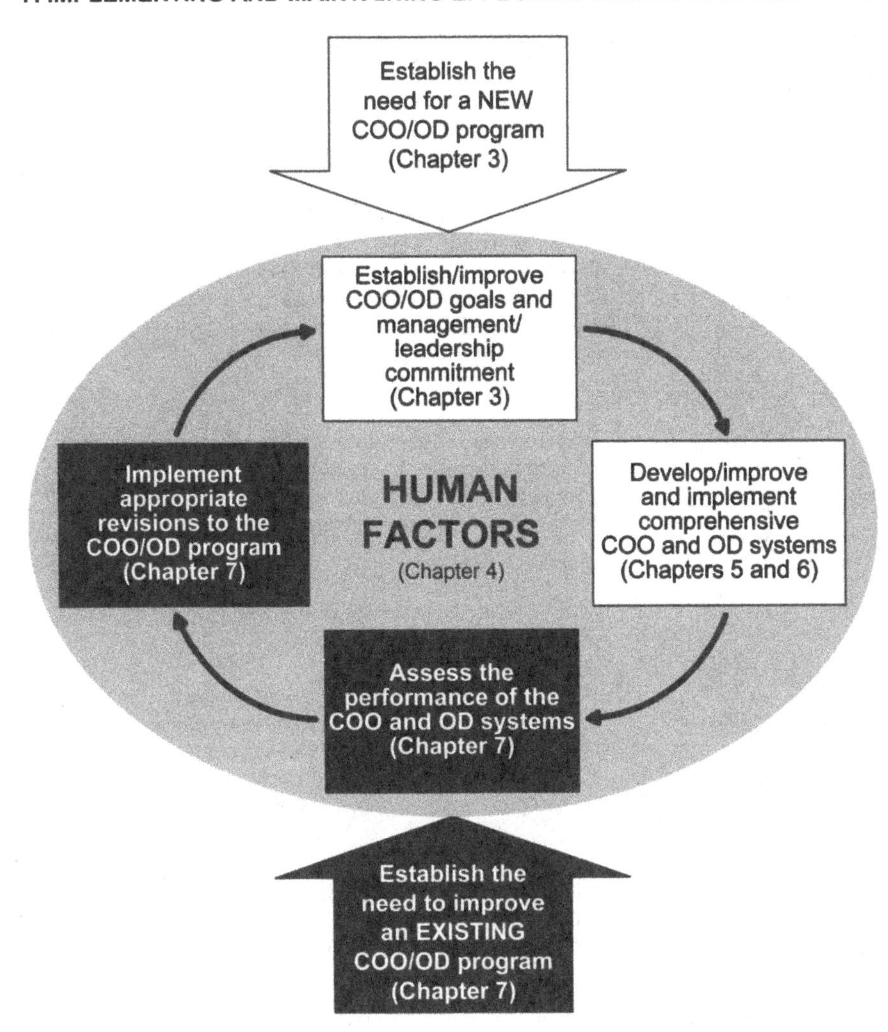

FIGURE 7.1. COO/OD Improvement and Implementation Cycle

Part of that reality is assessing the maturity, effectiveness, and compatibility of any existing management systems that address elements of COO/OD. This will help define the level of detail required in the plans to achieve the goals. For example, an organization will certainly have an existing training program, and it may be excellent. In that case, the planning activities will be relatively simple and may require nothing more than minor tweaking of the existing plan. An organization will also have existing shift-change practices, but perhaps intershift communications have been poor. In that case, a detailed plan will be required to address an immature element of the COO/OD system.

TABLE 7.1. The Plan-Do-Check-Adjust Process Applied to COO/OD Implementation

PLAN: Analyze the situation and develop a plan	Set a measurable objective toward the goal for the COO/OD effort Identify the processes impacted by COO/OD Select where to apply COO/OD List the steps in each process as it currently exists Map each process Identify issues related to COO/OD implementation Collect data on the current process Generate implementation plans Gain approval and support
DO: Implement the plan	Implement the chosen solution on a trial or pilot basis (first pass through the PDCA cycle) Implement the change throughout the organization (subsequent passes through the PDCA cycle)
CHECK: Evaluate the results	Gather data on the modified system results Analyze the results data Achieved the desired goal? ▪ If YES, skip the Adjust step, revise the goal to the next objective for continuous improvement, update the plan, and repeat the PDCA cycle ▪ If NO, proceed to the Adjust step, modify the implementation plan, and repeat the cycle
ADJUST: Standardize the implementation (and continuously improve)	Identify systemic changes and training needs for full implementation Plan ongoing monitoring of the COO/OD system Continue to look for incremental improvements to COO/OD

Mutual trust between managers and front-line workers is essential for the success of any improvement initiative. The only way that trust can be established and maintained is for all parties to clearly understand the standards to which they will be held accountable. In addition, there must be a mechanism to fairly judge performance against organizational standards and to hold people accountable for deviations from those standards. The organization must develop a strategic plan with realistic, achievable goals, set targets and timelines, and develop measurable performance standards.

7.2.1 Set Consistent Performance Expectations

The first element of the plan must be the establishment of consistent performance expectations, usually in the form of standards, throughout the organization. (For those seeking to improve or revitalize an existing COO/OD system with previously established performance expectations, this first step is a gap analysis as discussed in Section 7.5.1.) All employees, not just those from the operations department, will benefit from well-defined, well-developed, and well-communicated performance standards with clear lines of responsibility for both normal and emergency conditions. Developing effective performance standards requires cooperation between and consultation within every level of management, as well as the employees who will be affected by the standards. This will require an intense effort at each level in the organization to identify which tasks are (or should be) basic knowledge or job skills and which are controlled by specific procedures or guidelines. The process is conceptually similar to implementing an ISO 9000 program. (ISO 9000 is a family of standards for quality management maintained by the International Organization for Standardization). In fact, many of the procedures employed to deliver high-quality products and services can be used directly (or with slight modification) to *safely* deliver those same products and services.

In addition to establishing performance standards for critical tasks, the plan must address how the standards will be enforced. Who will draw the bright line between acceptable and unacceptable performance, and how will workers be held accountable? Many managers, perhaps unconsciously, believe that "bad people make bad errors," but the reality is that even the best people make serious mistakes. Respect and confidence play a vital role in the successful development, implementation, and maintenance of performance standards. Accountability is good and necessary for long-term success of any organization. People at every level want to know if they have performed well or not. They also want to be confident that management will not tolerate behavior that puts their personal safety at risk. By working together, the supervisor and employee can ensure that a fair and equitable performance standard has been set. The result is better productivity, a healthier workforce, and fewer management issues.

7.2.1.1 Characteristics of Performance Standards

The objective of any performance standard is successful task completion using the prescribed methods. Standards should be specifically tied to the worker's significant duties and responsibilities. The most effective performance standards are:

- **Measurable** and **observable**, so that both the employee and the supervisor can tell whether a task was or was not completed satisfactorily. Specific methods for collecting performance data and measuring performance against the standards should be identified.
- **Attainable,** so that any qualified, competent, and fully trained person who has the authority and resources can achieve the desired result.

Performance requirements within an employee's job duties should be fully within the employee's control to accomplish in terms of results (quantity, quality, time, cost, effect, etc.).

Each requirement of a standard should be evaluated to determine its purpose and what results or outcomes would be evident when performance meets expectations. Some terms for expressing performance standards are described below:

- Quality: specifies how well the task must be accomplished; specifies accuracy, precision, appearance, or effectiveness
- Quantity: describes how much work must be done
- Timeliness/rate: specifies the time frame within which the results are to be accomplished. A standard for expected timeliness can be as simple as "Completes assignments (projects, briefs, analyses, etc.) on schedule." Where the flow of work assignments is regular and predictable, the standard for expected performance can be more specific; for example, "Assignments are completed within a specified time (minutes, hours, weeks, etc.) after receipt" or "On average, ten transfers are completed during a shift."
- Effects of effort: describes the ultimate goal to be accomplished; expands statements of effectiveness by using terms like "in order to," "as shown by," or "so that"
- Effective use of resources: specifies how performance will be evaluated in terms of using resources (cost avoided, money saved, waste reduced, percent improvement, etc.)
- Methods of performing assignments: differentiates between situations in which only the officially approved procedure, policy, rule, regulation, or guideline for completing the task is acceptable versus those considered skill-of-the-craft
- Cooperation: describes situations in which an employee must work as part of a team to achieve the expected performance

In addition to clearly stating the requirements, well-written performance standards should have the following characteristics:

- **COO performance standards focus on behaviors and activities, not results.** When managers and supervisors focus on short-term results, it encourages behaviors that may be detrimental to the long-term success of the organization. A common misconception among managers and supervisors is that behaviors and activities are the same as results. Therefore, an employee following the prescribed procedure may not appear as productive as one taking shortcuts. The supervisor may mistakenly believe that the "more productive" employee is more dedicated to the job, more committed to the organization, and consequently deserves a higher-than-average performance score. However, from a COO perspective, exactly the opposite may be true because the worker is likely achieving those short-term results by using unapproved procedures and taking unacceptable risks. If there are superior ways to perform the task safely, then the procedure should be revised through a controlled MOC process. The training associated with procedures (or changes) should explain why certain steps are critical so that workers understand the importance of doing them the prescribed way.

- **COO performance standards facilitate meaningful measurements.** Measurements are helpful for benchmarking, standardization, or best practices for comparison to other organizations or to other parts of their own organization. They provide a consistent basis for comparison during the PDCA cycle. Performance standards indicate results during improvement efforts, such as employee training, management development, and quality programs. They also help ensure fair and equitable treatment to employees during the performance appraisal process. The best standards include both leading and lagging indicators as discussed in Section 7.4.1. For example, failure to perform required maintenance tasks in a timely manner (a leading indicator) should be a good predictor of unplanned shutdowns (a lagging indicator) that might not be evident for months or years.
- **COO performance standards align individual activities and processes to the goals of the organization.** Performance standards identify organizational goals, the results needed to achieve those goals, the measures of effectiveness or efficiency related to the goals, and the means to achieve the goals. This chain of measurements is examined to ensure alignment of individual activities and goals with the overall goals of the organization. This allows individuals to see how their actions contribute to the overall success of the organization.
- **COO performance standards support communication.** Performance standards not only support ongoing communication, they also provide dialogue and feedback regarding organizational goals and objectives. Communication between supervisors and employees can be framed within the context of objective performance standards rather than emotions and opinions.

7.2.1.2 Developing Performance Standards

Developing performance standards is a straightforward process of asking and answering why a particular activity is performed. Standards should be created for key areas of process safety responsibility and written into the employee's job description, particularly for individuals such as engineers and managers whose work activities may not be spelled out in specific procedures. Front-line workers will more willingly embrace OD to meet the standards if they are meaningfully involved in the development of those standards. Collaboration reinforces the team concept, empowers the employee, and promotes trust, which is fundamental for establishing and maintaining positive and progressive employee relations.

> **Example Performance Measures**
>
> Manager – provides development opportunities to subordinates
> Engineer – resolves design review issues on schedule
> Operator – completes assigned rounds
> Maintenance – completes inspections on schedule
> Purchasing – maintains required inventories

It is not necessary to create standards for every task; however, emphasis should be placed on those tasks that are most important to process safety. The standards should define operating objectives and the means to achieve those objectives, and should clearly define individual responsibilities. Performance standards should establish suitable benchmarks against which an employee's operational discipline in executing those responsibilities can be gauged. The following steps should be used to develop relevant performance standards:

1. Identify the employee's safety-critical job responsibilities.
   a. Review the job description to verify that it reflects the position's current responsibilities.
   b. Identify the particular duties on which the employee expends most of his or her time.
   c. Identify the fundamental activities required to complete each job responsibility.
2. Identify the measurable and objective performance standards.
   a. Identify performance indicators for the three to five key job responsibilities.
   b. Write these performance indicators as specific, observable, measureable standards.
   c. Evaluate how the performance indicators correlate to the desired results.
   d. Define the absolute minimum acceptable level of performance.
3. Identify the means by which the employee's operational discipline will be monitored.
   a. Evaluate the employee's final work output.
   b. Examine and observe the manner in which the employee performs each duty.
   c. Review status/progress reports, feedback from customers/clients, and other records and documentation.

As management's first-line representative in an organization, the supervisor has specific responsibilities for the development, implementation, communication, and maintenance of COO performance standards. In preparing for the development effort, the supervisor should first determine that the appropriate methods of operation (standard operating procedures) are being followed by the worker. This includes tasks such as the following:

- Confirm that the workplace or workspace is logically arranged and conducive to the required tasks. The materials, tools, equipment, and controls should be located to maximize productivity. While it may take years to resolve all of the physical impediments that may be present in an existing facility, often small improvements in the layout of the workspace will correct the most significant issues without incurring large costs. Nevertheless, an inefficient physical arrangement does not excuse shortcuts or other breakdowns in OD.

- Verify that the equipment is set up properly and is operating within the appropriate specifications, tolerances, and safety parameters. A thorough check should be made to ensure that the proper inspection and safety equipment (e.g., H_2S detectors, lower explosive limit detectors, fire extinguishers) is in place and that the employee is familiar with its use.
- Ask the employee to describe the safest and most appropriate way of using equipment and performing both normal and emergency job functions (e.g., starting pumps, breaking flanges, taking samples, containing spills, evacuating).
- Observe enough individuals performing the job tasks to ensure that the operation can be consistently performed to produce the desired results.

In a COO/OD system, the performance standard typically addresses topics such as the following:

- **Procedures.** Who is responsible for identifying the required procedures? Who is responsible for developing, validating, and maintaining them? Is there a standard for writing procedures and must all procedures be written (or converted) to that standard? Is there a mechanism for user feedback to be solicited and incorporated into the procedures? Are all procedures to be treated equally, or are some procedures more important than others? For example, any procedure classified as "critical" must include a checklist that is signed off, step-by-step, each time the task is executed. Some procedures may be required at the work location to aid in performing the task, although a step-by-step check-off is not required. Other procedures might simply be available as reference when the worker is preparing for or executing the task, at the worker's discretion.

> An operator had three different procedures for a simple cleaning operation. Some instructions conflicted, and there were gaps between others. The outcome was a flash fire that resulted in minor burns to the operator. Had the cleaning procedures been reviewed with the operators, the deficiencies could have been corrected. However, cleaning procedures did not warrant annual review like their "production" counterparts.

- **Training.** Who is responsible for deciding what training is required, as well as for developing and delivering the required training? Will refresher training be required and, if so, at what frequency? Is there a standard for developing training courses and must all courses be written (or converted) to that standard? Is there a standard for developing instructors (e.g., lecturers, on-the-job trainers, mentors) and must all instructors be qualified to that standard (e.g., technical knowledge, presentation skills, coaching ability)? How will student progress be measured (e.g., by written tests, field demonstration, or apprenticeship) and what level of achievement is required to successfully complete the training?

- **Staffing.** What are the minimum staffing requirements for each phase of operation (e.g., startup, normal operation, shutdown, maintenance) and what must be done (or what cannot be done) if the minimum staff is not present? To what extent can overtime be used to fill staffing shortfalls? Under what conditions can personnel from other units be "borrowed" to complete tasks?
- **Ownership.** What work group "owns" the equipment at any given time? How is ownership transferred from one work group to another? For example, how much overlap is required at shift change and what information must be communicated in written form? Under what circumstances are field walkdowns of the work location required? Under what circumstances does maintenance or engineering assume ownership and how is it returned to operations? Who is responsible for troubleshooting and problem solving?
- **Verification.** What is the expectation for workers to verify the condition of equipment during normal use or operation? Is risk a consideration? For example, is it sufficient for the crane operator to verify the condition of his or her equipment at the beginning of the shift, or must verifications be performed before each lift? Are special verifications required if there is a change in equipment ownership, such as the transfer from engineering to operations after equipment has been installed or modified?
- **Communication.** What is the expectation for written and verbal communication within and between work groups and shifts? Is verbal communication alone acceptable, or must it be accompanied by written confirmation? Do communications require confirmation and/or acknowledgement, such as repeat-back of verbal instructions? If a worker simply opens an e-mail, is that sufficient confirmation that the message has been received and understood? Who is responsible for posting and maintaining labels and signage?

7.2.2 Focus on Management Leadership and Commitment

Successful implementation will depend on a combination of factors, but sustained management leadership and commitment are foremost among them. Therefore, in addition to performance standards, the plan must include a strategy for engaging all levels of management. Just as upper managers must model the COO/OD behaviors they expect from the organization, subordinate managers (down to the front-line supervisors) must also model the desired behaviors so that all employees feel their influence. (DuPont uses the term "felt leadership" [Ref. 7.2]). As further discussed in Section 7.4.4.1, nothing facilitates open communication more than visible leadership – managers in the plant interacting with the front-line workers. Each level of management should reinforce the organization's vision and exhibit solid support for COO as the means to improve safety and quality.

COO must become part of the organization's culture. In other words, COO must be embodied as a fundamental element in the organization's rituals, traditions, and activities. Leadership must understand the business case for COO and articulate a compelling case so that everyone understands and shares the goals. Each level of management should have a defined role in communicating the goals, policies, and standards of the COO system.

> In the 1870s, Dr. Theodor Billroth was one of the first physicians to introduce antiseptic surgical practices. Through his leadership, the hospital's organization/ infrastructure was changed to an integrated system. Dr. Billroth was able to get his practices adopted widely, and he had much greater actual impact on reducing mortality rates than other, more famous physicians of his day who adopted antiseptic practices individually.

Active communication is the cornerstone of success and, by definition, it is a two-way process of both speaking and listening. Through active communication, management helps build and maintain a common set of shared values, such as safety as a core value being on equal footing with financial performance. Through active communication, workers share their concerns and ideas for improving the COO system. This ongoing dialog shows management's enduring commitment to COO/OD and helps bridge any gaps in workers' perceptions of management's support for COO/OD as an avenue to a safe and productive workplace. This will also reinforce the systemwide alignment of strategic initiatives into one overarching strategy; that is, integration of safety into every aspect of operations.

Active communication also engages the workers in reporting COO/OD activities so that successes can be shared with others and deficiencies can be addressed. Progress towards goals should be reported at every organizational level through a dashboard of key metrics. This will promote accountability and reinforce incentives to meet both target and stretch goals. Management support includes rewards (financial and otherwise) for higher performance and implementation of COO.

The plan should also elicit active worker involvement and leadership. Management can engage workers in COO working groups to develop strategies and policies for implementing COO/OD across the organization. Management then supports the committees by providing staff and resources, participating in collaborative meetings, and completing actions, such as the redesign of work processes per worker recommendations. Thus, operational excellence is operationally, rather than administratively, driven. The objective is to empower the workforce within the COO/OD system.

Another key role of management, as previously discussed in Chapter 3, is to assign people to roles in which they can be successful. Thus, a staffing plan must be developed so that managers and front-line workers who embrace COO/OD principles are placed in positions of authority and leadership. They can then help build and lead a culture of personal accountability for safety and operational discipline just as Dr. Theodor Billroth influenced others to improve medical practices. In addition, the plan must address how to attract and retain people of similar mindset. This should include clear incentives and positive reinforcement, but does not necessarily require significant resources. In fact, many of the rewards for disciplined behavior – a personal sense of accomplishment, team membership, respect, etc. – are intangible.

7.2.3 Focus on Long-Term Sustainability and Consistency

A COO/OD system is a program, not a project. A project is a set of events with a beginning and an end for delivering a unique product, service, or result. Project objectives define the target for the project. On the other hand, a program is defined as a group of related projects managed in a coordinated way to obtain benefits and control not available from managing them individually. A program is an ongoing collection of initiatives and projects that are designed to accomplish a strategic business objective.

While activities associated with COO/OD development and implementation may be organized as a project, the emphasis should be on long-term sustainability and consistency. Its benefits will derive from long-term improvements to organizational culture and behavior. The scope is relatively flexible and will likely change over the years with the changing needs of the organization and progress toward its goals. At any given moment, there may be disagreement between stakeholders on the preferred path forward. The risk of failure is real, and the potential impact on the organization could be great. Therefore, it is imperative that the problem be clearly defined, the objectives be SMART, and the immediate scope be narrow and tightly defined. However, the reality of the organization's current situation is dynamic, so the program's objectives need to be managed within the context of the changing organizational environment.

Behaviors promoting safety must be embedded in operations and include simplified, standardized processes and the elimination of non-value-added activities. The focus is on risk reduction, with the ultimate goal of eliminating workplace injuries and illnesses while having the least impact on production. Metrics should be selected so that any boom-and-bust cycles are evident. For example, if a housekeeping milestone is achieved by a frantic weekend of overtime cleaning, then the COO objective of maintaining good housekeeping is not yet ingrained in the work processes – even if the housekeeping inspection shows exceptionally good results.

For long-term success, the plan should focus on developing effective in-house processes and embedding them into the organization's daily operations. When incidents do occur, root cause analysis will provide feedback on aspects of the management system that should be improved.

7.2.4 Set a Few Milestones and Push to Achieve Them

When first implementing a COO/OD system, there is an overwhelming temptation to try to do too much, too soon. Opportunities for improvement abound, and managers want to chase every opportunity. Unfortunately, the chaos resulting from such an unfocused strategy will usually result in failure.

A more productive approach mimics the lioness in her hunt for food. She ignores most of the herd and focuses her efforts on a few prime targets with the highest probability of falling prey. She has the same core goal (to eat) every day, but her immediate focus is on the most opportune target, given today's circumstances. Like the lioness, implementation teams need early successes; however, they must start with a long-term strategy and maintain their focus on building a solid foundation.

Teams implementing a COO/OD system should focus their efforts on milestones that are reasonably achievable, given the organization's current circumstances. Milestones should be challenging, and progress toward them should be measurable and auditable. However, milestones should be set as motivators for improvement, not as ends in and of themselves. Attaining a milestone should move the organization toward attaining its core goals and give everyone involved a sense of accomplishment. Milestones that can be met with little or no effort or that fail to provide meaningful benefits are useless. While celebrating each success, management can then slightly shift the organization's focus to the next milestone as a new challenge for further improvement. Thus, the COO/OD system is continuously improved by setting and achieving many incremental goals.

> On December 14, 2004, Dr. Don Berwick, the chief executive officer of the Institute for Healthcare Improvement, specifically challenged the medical profession to move beyond vague goals for improving patient safety ("some is not a number, soon is not a time"). He announced a campaign to save 100,000 lives in the next 18 months. The effort's structure and focus were modeled on a political campaign. Conceptually, the idea of a campaign catalyzed by front-line workers "doing the right thing" was attractive, and it was largely successful. (Ref. 7.3)

The plan should include promotional strategies to enhance COO/OD awareness among the employees and drive ownership as close to the front-line workers as possible. Promotional strategies could include the following:

- Mission statements, slogans, and logos
- Published materials (library, statistics, newsletters)
- Media (posters, displays, audiovisuals, e-mail, Internet)

Training and awareness activities should include short talks; group meetings; training on COO principles (including compliance with rules and regulations); participation in hazard identification analyses, risk assessments, and incident investigations; and job safety analyses. These activities should be designed to allay employees' concerns, clarify expectations, and address the proverbial question,

"What's in it for me?" Promotions might also include special campaigns such as Housekeeping Week or Human-Machine Interface audits. When employees become more aware of their COO responsibilities for incident and injury prevention, they will exhibit more interest in maintaining a safe and healthy worksite.

The challenge, then, is to design, implement, and propagate tools that will allow organizations to monitor their commitment to COO/OD, modify business processes, achieve transformation, and reach the next milestone in organizational performance.

7.3 IMPLEMENT THE PLAN

The second step in the PDCA cycle is the most difficult because it truly tests whether management has the courage of its convictions. The time for discussion is past – management must steadfastly press forward to do what it deemed necessary to achieve the organization's goal. Change inherently requires a leap of faith that the long-term results will be worth the near-term costs and risks. Change often results in a short-term decline in performance before a higher level of performance is attained. Thus, consistent and persistent management leadership will be required to convince others of the necessity and benefits of change, even in the face of a short-term decline. Management must effectively communicate new performance expectations, provide the necessary resources, and manage the changes in process flow, job function, tasks, and activities that will be required to meet the new performance standards. Management must also be prepared to adapt the implementation plan to site-specific realities without compromising the core values.

Once new performance standards have been established, management must enforce them. Issues that occur during the transition to new performance standards normally arise because of poor communication across the organization. When there is confusion regarding performance standards and employees ask questions about them, the supervisor must be able to address these concerns quickly. Thus, supervisors must be involved in developing the standards and must believe in their value so that they can speak persuasively. By resolving questions promptly and accurately, supervisors earn the trust and respect of their employees and eliminate issues before they become major problems.

7.3.1 Start with the Benefits – What's in It for the Workers?

In promoting COO/OD, management should emphasize that the goal of a safer workplace is the responsibility of every employee. For a positive COO/OD culture, employees' involvement, ownership, and commitment are necessary; empowerment promotes feelings of self-worth, belonging, and value. The obvious long-term benefit of COO/OD is that it enables everyone to work without injury so that they can continue to provide earnings for both themselves and the company. Management must instill an across-the-board belief in the organization that workplace injuries and illnesses are avoidable and that they can be eliminated through application of COO/OD.

However, worker behavior is more strongly influenced by near-term benefits. Employees will see some immediate benefits in improvements that make the workplace more pleasant (e.g., housekeeping) and in improvements that make their jobs easier (e.g., clearer procedures, less rework). In addition, most organizations have programs that reward workers financially if economic goals are met. However, if the rewards are simply tied to business results, then some employees may try to achieve those rewards by cutting corners, not observing safety rules, not wearing personal protective equipment, and ultimately not working safely. Thus, for any financial incentives (promotions, raises, and/or bonuses) to succeed, they must be linked to improved performance as measured by COO/OD metrics as well as improved economic performance.

Joseph Scanlon was a labor leader during the 1930s, and his concept of "gainsharing" is often used as a method to drive organizational change. Gainsharing is much more than a simple bonus system because it includes a structured system for worker involvement and requires collaboration between labor and management. It combines metrics from several key performance areas (finance, COO/OD, safety, environment, quality, etc.) in a formula to calculate the size of a monetary bonus pool that is shared among the workers, managers, and owners.

7.3.2 Communicate Performance Standards

Performance standards describe the conditions or the quality level that must exist before performance can be rated as acceptable. The key to effective performance is communicating expectations, and one way of doing so is through performance standards. Employees are more prone to excel when they clearly know and understand what is expected of them.

In communicating the performance standards, management should focus on the individual worker's duties and responsibilities in completing the task. Standards should be communicated to the workers in their training programs, and supervisors should communicate standards in their everyday monitoring and guidance of work activities. The essence of COO/OD is doing the job correctly the first time, every time, to achieve the organization's goals.

Management should use performance standards as a basis for discussion during periodic coaching sessions with employees. This reduces ambiguity and allows the opportunity for more objectivity when providing feedback throughout the year, as well as during the annual performance appraisal process. Referring to performance standards will be beneficial when the manager or supervisor is describing gaps between acceptable and unacceptable performance.

When communicating new performance standards, management should be prepared to address worker resistance to change, but it should not assume that the workers will resist performance improvement. Perhaps past initiatives simply lacked a clearly articulated, meaningful opportunity for workers to do something. COO gives workers a specific way to achieve an organizational goal, and management may be pleasantly surprised to find a wellspring of energy from workers who will eagerly commit themselves to the cause.

7.3.3 Implement and Enforce Performance Standards

Once the standards have been developed, senior management must ensure that everyone (front-line workers, supervisors, and managers) is well trained on the COO/OD system. Workers should clearly understand their authority, responsibility, and required interfaces with other work groups. Workers must acknowledge their individual responsibility to know their job requirements and personal capabilities, to execute tasks correctly and safely each time, and to seek supplemental or refresher training as needed. Then management must actively implement and enforce the performance standards. Management should provide the necessary resources, closely monitor performance, and look for opportunities to reinforce desired behaviors.

Behavior is reinforced when it is explicitly tied to positive consequences for the person. The sooner and more certain the positive consequences are received, the more strongly the behavior is reinforced. Thus, reinforcement of new, desired behaviors should occur whenever the behavior occurs. Reinforcement can be a simple compliment or sincere "thank you" from the supervisor, and over time most workers will internalize the reinforcement of desired behavior as personal satisfaction for a job well done. As more workers exhibit the desired behaviors, the cumulative effect should produce positive results for the organization. When defined thresholds are met, management should use those as opportunities to reward the workers for their progress and celebrate individual heroes who exceeded expectations. However, routine positive reinforcement is a much stronger influence on behavior than occasional rewards.

Management should also anticipate resistance and challenges to the COO/OD system, partly due to humans' natural resistance to change and partly to simply test management's commitment to the new system. The design of the plant will also influence workers' willingness to implement and maintain COO/OD. Nevertheless, a central tenet of OD is that workers at every level will be held accountable for their performance. A mixture of supervisory counseling, performance appraisals, and positive reinforcement will overcome most initial implementation problems, but some remedial training may be required.

More serious resistance may require progressive disciplinary measures, so there must be a plan for dealing with such cases in conjunction with the human resources department. In particular, supervisors should be alert for personality characteristics that will conflict with the adoption of COO/OD, such as:

- "Above the rules" mentality
- Risk-taking behaviors
- Willingness to ignore rules or violate codes/laws/regulations
- Uncooperative behavior
- Attempting tasks beyond one's competence
- Attempting to be a hero

Personnel involved in frequent violations of operating practices should be counseled, retrained, and disciplined, as appropriate.

At any given moment, management is getting precisely the performance that the organization is tuned to produce. To change the status quo, management must be willing to change itself, the facility, the procedures, and ultimately the workers as necessary to implement and achieve the new levels of performance.

7.3.4 Adapt the Approach to Site-Specific Conditions

As discussed in Section 3.3.8, it is foolhardy to expect that a generic plan of any sort could address the complex and idiosyncratic issues of a multinational, multicultural, and/or multiregional workforce. Workers may agree with the objectives of COO/OD, but they will have different perspectives on what makes a good procedure, how best to resolve conflict, and what is important in getting the job done.

Thus, the implementation plan cannot be a one-size-fits-all approach. Implementation guidelines must be flexible so that they can accommodate everything from a large production facility to a small pilot plant. Facilities may use approaches or methods other than those defined in the corporate guidance, but facilities everywhere are expected to meet the intent of the guidelines.

> Performance dashboards can create their own problems if the indicators are not comparably important. Management may become so obsessed with correcting a minor indicator "in the red" that it fails to address a far more important issue "in the yellow."

Local management must determine where the site is with respect to the COO/OD system. The simplest approach is to take the corporate plan as the baseline and jump forward to the **Check** step in the PDCA cycle. Management can then use any of the assessment tools discussed in Section 7.4 to determine where the local facility is better than, worse than, or simply different from the baseline implementation plan. Management should not merely compare a facility's written performance standards against the plan; the gap analysis should be based on real-world data to define the starting point for COO/OD implementation at a specific facility. There can be substantial and surprising differences in the actual implementation of existing standards. The plan can then be adjusted accordingly to begin the PDCA cycle.

7.4 MONITOR PROGRESS

The purpose of implementing a COO/OD system is to move the organization toward its goals. The **Check** element of the PDCA cycle provides the objective measures by which progress can be monitored and judged. Thus, it is essential that the plan identify such measures and that the supporting data be collected. Inspections, audits, management reviews, investigations, and self-assessments are all means by which data can be collected, as illustrated in Figure 7.2. Some metrics, such as the number of late/missed inspections, may be collected directly; others may result from other management systems, such as incident investigations and audits.

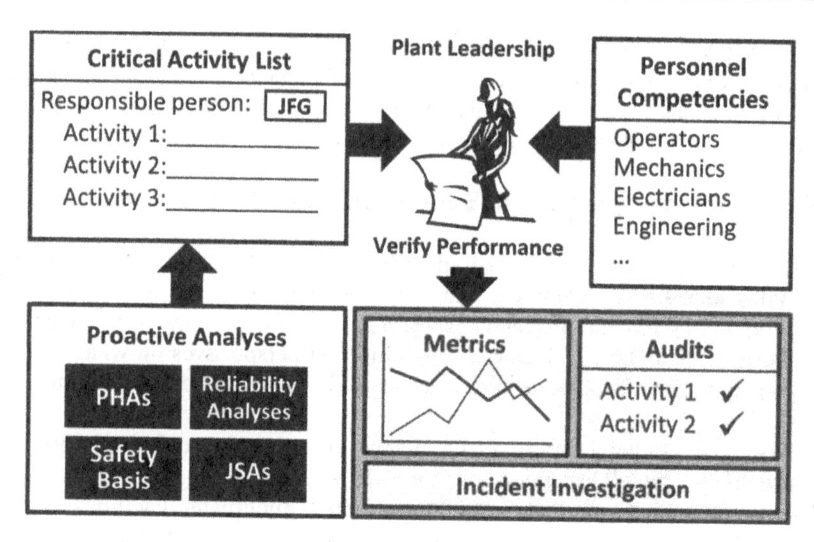

FIGURE 7.2. Monitoring Performance of Safety-Critical Tasks

A robust information system infrastructure focused on quality and safety is essential to the **Check** step. Systems that support that framework include digital process archives, activity tracking boards, checklists, a dashboard of metrics on process operations and safety, computerized maintenance and inspection records, and data mining for regular reports of progress on key safety and quality indicators. To be useful these processes must ensure data accuracy and timeliness.

Management can advance the COO/OD system by being transparent when reporting metrics on system performance. For example, a company may exhibit good COO/OD performance in meeting quality goals but poor COO/OD performance in meeting safety goals. Everyone should be able to see the organization's areas of growth and success as well as those areas that need more attention and improvement. Sharing this information is important in transforming the organization, and it will improve workers' ability to understand how COO/OD impacts every aspect of the business. To that end, a system using a balanced scorecard will achieve a better alignment between strategy and daily activities and better communicate strategy and performance throughout the organization. This scorecard ensures a "balanced" focus on behavior and results, and it may consist of multiple layers of cascaded scorecards, which ultimately align individual facility and business unit results with the key top-level strategic objectives and performance measures for the entire organization.

The benefits of the balanced scorecard system include the ability to (1) communicate strategy to all organizational levels, (2) provide alignment and "drill down" visibility into strategic objectives, metrics, and problem areas, (3) ensure that resources are applied appropriately, and (4) drive accountability and results. Standardizing the terminology will increase the value of the scorecard as a communication tool.

7.4.1 Use of Metrics

As discussed in Section 3.3.4, metrics are performance and efficiency indicators that allow workers and managers to monitor the near-real-time effectiveness of the COO/OD system and identify where improvements are needed.

Ideally, the chosen metrics would reliably warn of impending consequences and allow fair comparisons between operating units or plants; however, such reliable predictors are rare in the real world. Most of the leading indicators are activity measures, such as the number of inspections completed or attendance at safety meetings, and their predictive ability is asymmetric. For example, a declining number of field inspections may strongly warn of impending failure, but a constant or increasing number of inspections may only weakly correlate to continued success. Metrics can even drive undesirable behavior, such as a worker neglecting other equally important but untracked duties to make time for more field inspections, which are being tracked. Or workers may simply not report an incident that would adversely affect a metric. When metrics are used, it should be stressed to everyone that they will not be punished for reporting incidents that may degrade the indicator.

> An organization had a mechanical integrity metric that was based on the percentage of on-time inspections. One plant was praised for 98% attainment while another was criticized for 88% attainment. In reality, the "worse" plant was not reporting the inspection as complete until the results were documented in the mechanical integrity database, while the "better" plant was reporting inspections as complete when the field work was completed. When this discrepancy was resolved, the higher-scoring plant's performance was actually below 88%. The discrepancy highlighted that the measures were not clearly defined and were open to inconsistent interpretation.

Thus, a blend of leading and lagging indicators is the best way to provide a complete picture of COO/OD effectiveness (Refs. 7.4, 7.5, and 7.6), and some of the metrics should be changed periodically to provide a new perspective on the workings of the COO/OD system. In most organizations, COO/OD metrics can be based on data already being gathered for other purposes; new data-collection efforts are not usually necessary. Existing data can be supplemented, as needed, with brief data-collection campaigns to answer specific questions about COO/OD performance.

Some example metrics are listed below with a description of the COO/OD characteristic for which they might be useful in gauging (a more extensive set of metrics is included in the accompanying online material). The selection of specific metrics should follow the standard PDCA approach: (1) decide what metrics would be useful and how to collect them, (2) collect the metrics for a period of time, (3) check that the metrics provide useful information at reasonable cost, and (4) adjust the metric-collection effort as appropriate. In many cases, COO/OD data can be gleaned from existing metrics (e.g., incident reports). Every situation is different, so management will need to determine how to track and present data in a manner that best serves the need to monitor the health of the COO/OD system in its current state at their facility.

- **Frequency of communicating progress toward goals.** A declining communication frequency may indicate declining management interest or a reluctance to confront bad news. Declines also correlate with lower worker morale because they see no progress as a result of their efforts.
- **Number of audit findings related to inoperable instruments and tools.** Increases may indicate that operators are not attentive to the human-machine interface or that management is not providing resources to promptly repair deficiencies.
- **Number of housekeeping audits and their scores.** Decreases in the number of audits and declining audit scores may indicate declining interest by management and workers. A stable or increasing number of audits and stable or improving scores normally indicate that good housekeeping is ingrained as part of operating practices.
- **Percentage of overdue corrective actions.** Corrective actions may be generated by audits, incident investigations, process hazard analyses, etc. An upward trend or unexpected spike may indicate poor OD by those responsible in management, engineering, maintenance, or operations.
- **Average time to resolve off-normal situations or findings.** An increase may indicate inadequate resources or an increasing tolerance for process deviations. (Note: Considering risk enhances the value of this metric, which could be measured in risk-days. For example, a high-risk issue could be weighted 9, a medium-risk issue weighted 4, and a low-risk issue weighted 1. Resolving a high-risk issue in three days is much more important than resolving a low-risk issue in one day.)
- **Incidence of shortcuts identified by near misses and incidents.** Increases in these numbers may indicate failure to enforce best practices, overly aggressive performance goals, inadequate staffing, or rewards favoring results over behavior.
- **Number of safety, environmental, production, or quality incidents with OD as a key factor.** An increase may indicate deficiencies in the COO/OD system, deterioration in workers' understanding of OD, or deficiencies in management's communication and enforcement of OD expectations.
- **Number of incomplete shift logs, reports, or turnovers.** An increase may indicate a lack of discipline or excessive operator workloads or distractions.
- **Shift production rates that exceed expectations.** An upward trend may indicate that workers are taking shortcuts or maintaining inadequate safety margins.
- **Unit uptime or yield factors are below target values.** The difference could result from a variety of OD problems, such as poor inventory management, scheduling, equipment reliability, or adherence to procedures.

- **Number of nuisance and always-on alarms.** An increase may indicate excessive operator workloads or distractions, poorly defined operating limits, or inadequate resources to repair faulty devices.
- **Unavailability of safety systems.** An increase may indicate an increasing tolerance for process deviations or inadequate resources to repair faulty devices.

7.4.2 Use of Audit Results

Audit findings of incomplete or overdue work activities show a breakdown in operational discipline:

- Overdue inspections
- Overdue training
- Out-of-date procedures
- Out-of-date drawings
- Incomplete work permits
- Late reports to regulators

An audit is a systematic, independent review to objectively verify conformance with prescribed standards of care. It employs a well-defined review process to ensure consistency and allow the auditor to reach defensible conclusions. Audits should be conducted throughout the development and implementation of the COO/OD system, and periodically thereafter. The nature and frequency of the audits will be governed by factors such as the current life-cycle stage of the facility, the maturity (degree of implementation) of the COO/OD system, past experience (e.g., prior safety performance and audit results), and applicable facility or corporate requirements.

Virtually any audit can provide useful information as to whether COO/OD systems are performing as intended. Audits complement other control and monitoring activities such as management reviews. However, typical PSM audits do not address the full range of COO/OD issues. Therefore, audits specifically targeting COO/OD topics are essential to ensure effective, consistent implementation of the COO/OD system throughout the organization and to verify the integrity of the metric data collected (Ref. 7.7). Topics included in a COO/OD audit might include:

- COO/OD system development, quality, and status
- COO/OD training and expectations
- Management visibility and leadership by example
- Worker knowledge, commitment, and awareness
- COO/OD metrics for each element (communication, housekeeping, work permits, etc.)

While they can be scheduled as needed, audits should be conducted at some predetermined interval; frequencies ranging from once per year to once every three years are common. Data are gathered through the review of documentation and implementation records, direct observations of conditions and activities, and interviews with individuals having responsibilities for implementation or oversight of the COO/OD system or who might be affected by it. The data are analyzed to assess compliance with requirements, and the conclusions are documented in a written report.

A finding is a conclusion reached by an auditor based upon data collected and analyzed during the audit. Findings indicate a deficiency in the implementation of the COO/OD system based on the organization's current requirements. Findings must be resolved during the **Adjust** phase of the PDCA cycle. An observation is a conclusion reached by the auditor that is not directly related to compliance with the standard of performance. Observations may be good COO procedures or practices identified during the audit that should be shared across the organization. Observations may also result when the auditor believes that, while the requirements established by the standard(s) have been met, opportunities remain for improving the implementation of the COO/OD system. Like findings, observations should be addressed during the **Adjust** phase of the PDCA cycle.

> **Example finding:** Written shift logs were not kept as required.
>
> **Example observation:** Summarizing the shift log on a standard form improves communication between shifts.

Audit results should be trended over time to determine whether COO/OD performance is improving, with adjustments made as necessary. Repeat findings are particularly worrisome because they indicate that corrective actions were ineffective.

7.4.3 Use of Incident Investigations

Serious process safety incidents, when they do occur, usually involve a confluence of root causes, some of which may involve a degraded COO/OD system. Thus, the data gleaned from incident investigations offer unique insight into specific weaknesses in the COO/OD system. Incident investigation data should also be trended to spot the recurrence or confirm the resolution of past problems.

Because OD includes an element of accountability, some managers mistakenly use the incident investigation process to assign blame to individuals involved in an incident. This approach is always a mistake because it simply drives the reporting of minor incidents and near misses underground, so there is no opportunity for organizational learning. OD applies to everyday activities *before* an incident occurs. After an incident occurs, the value of learning how the management system(s) failed far outweighs the value of punishing individuals. Simply counseling an employee to "be more attentive" is unlikely to resolve an OD problem. A better approach is to develop effective recommendations and assign them to individuals who will be held accountable for correcting the underlying, system-related causes of incidents. A serious incident may identify focus areas for immediate adjustment in the fourth step of the PDCA cycle.

Thorough incident investigations usually find deficiencies in both COO and OD. For example:

- The operator failed to follow procedure (OD), but the procedure was out of date (COO).
- Thickness inspections had not been performed (OD), but the department was understaffed (COO).
- Drawings were not reviewed (OD), but the procedure did not address changes by a vendor (COO).

Incident investigation is a way of learning from incidents that occur over the life of a facility and communicating the lessons learned to both internal personnel and other stakeholders. Depending upon the depth of the analysis, this feedback can apply to the specific incident under investigation or a group of incidents sharing similar root causes at one or more facilities. Leaders must set the tone, learn how to listen, and talk about COO/OD concerns continuously – with front-line staff and at the highest levels of the organization.

The goal is to reduce the number and severity of process safety incidents, and this can best be accomplished in an atmosphere of open communication. Organizationally, leaders must put in place an interdisciplinary review process so that when an error occurs, everyone involved – all those on the front line – sit around a table and talk about the experience. Managers must be there to support the front-line workers (who may be devastated by an accident) and ensure that the findings are addressed.

When evaluating incident data, managers should be wary of concluding that if there are no events, there are no human performance problems. Human errors occur every day. The absence of events is more a function of the presence and integrity of defenses, barriers, controls, and safeguards than the errors people make. Or the organization may simply have a "nonreporting" culture. Therefore, it is erroneous to believe that human performance is adequate just because the facility has not experienced significant events. Incident investigations may not have been required, so management must rely on metrics, as previously described, to gauge the health of the COO/OD system.

7.4.4 Use of Other Tools

7.4.4.1 Routine Interaction with Facility Personnel – Management by Walking Around

One of the best approaches to monitoring the COO/OD system is for members of management to simply walk around the facility periodically. It is easy to believe that everything is okay simply because production numbers are good and there are no reports of serious incidents. Managers should proactively go looking for trouble, probe the staff, and ask people "What are your concerns?" Supervisors should encourage the reporting of "bad news" as well as success stories. Most workers want to do a good job and are eager to share their concerns, particularly if management is willing to help them overcome obstacles to success.

Management must provide opportunities for conversations that include feedback on performance, whether good or bad. Many COO/OD issues, such as poor housekeeping, poor signage, poor lighting, and poor communications, will be immediately apparent and can be discussed with workers in the area. Other issues will become apparent in discussions about what the worker is doing, why the worker is doing it, and what procedures and permits are governing the activity. In particular, managers should be alert to deviations in the desired performance because (1) that is how the worker routinely does it and there have been no problems, (2) that is the way others do it, (3) that is the way they were shown to do it, or (4) that is someone else's responsibility. These are opportunities for management to clarify the intentions and expectations of COO/OD and help improve front-line worker performance. However, management must avoid giving the workers the impression that they are "looking over their shoulder" trying to catch them doing something wrong. Even if everything is fine, the mere fact that management took the time to visit the shop floor, showed an interest in the workers' concerns, and acted to improve the situation will reinforce the importance of COO/OD and help ensure continued good performance.

7.4.4.2 Management Review

There are many specific questions/discussion topics that management will want to check periodically to ensure that the COO/OD system is being implemented and working properly. If the organization's performance is less than satisfactory or it is not improving as a result of management system changes, then management should identify possible corrective actions and pursue them. It is possible that the organization is not working on the right activities or that the organization is not performing the necessary activities well. Even if the results are satisfactory, are resources being wasted, or are there tasks that could be performed more efficiently or not at all? To help answer these questions, management can combine the metrics listed in the previous section with personal observations, direct questioning, audit results, and feedback on issues, as the following examples show:

- Discuss roles with workers to verify their understanding of their responsibilities and the lines of authority.
- Discuss possible upsets and incidents (e.g., via table-top exercises) with workers to verify their understanding of notification responsibilities.
- Discuss performance goals and current plant performance with operators to verify their understanding.
- Verify that current practices match policies and expectations; for example:
    - Radio traffic is free of idle chatter and nonstandard communication language.
    - Shift turnover logs are being kept contemporaneously and are transferred in an organized manner.
    - Tamper-indicating seals, drain plugs, and hatch covers are in place.
    - Pressure between rupture disks and relief valves is being checked routinely.

  - Work areas are free of distractions or unauthorized entertainment devices.
- Review the number of overtime hours worked by individuals and departments to determine whether there are adequate resources to perform necessary tasks.
- Discuss working conditions to determine whether shortcuts are being used to get the job done in time.
- Determine whether required cross-checks are actually being performed or are simply signed off.
- Check maintenance work orders to determine the percentage of "emergency" work.
- Review unit logs to verify that work groups are coordinating activities with responsible operators.
- Review the process for authorizing nonroutine activities for evidence of complacency.
- Monitor the number of visitors and administrative duties that distract operators from their primary tasks.
- Identify the number of nuisance and always-on alarms.
- Investigate whether maintenance is being deferred to meet production goals.
- Tour the work area to assess housekeeping and the status of required safety devices (e.g., chocks at truck stations, locks on critical valves, fire extinguishers charged and available).
- Investigate the reasons for any significant differences in the performance of different shifts, teams, areas, or departments.
- Determine how often supervisors observe work in the field.
- Determine whether peers are observing and coaching peers on an ongoing basis outside of formal training settings.
- Verify that the organizational chart is up to date and that clear responsibilities and lines of authority are being maintained.

Regardless of the questions that are asked, the management review should try to evaluate the organization's depth on several levels. Is the operational discipline sufficient? Does the formality of operations extend beyond routine tasks? When things go wrong, is it likely that key personnel understand the chain of command that must be followed to make prudent decisions? Also, is there sufficient depth in key personnel? If the entire organization depends on a single individual to make all of the really tough risk judgments, what will happen if that person suddenly resigns or falls ill? Management reviews provide an opportunity for the organization to honestly assess its depth, and to take action to address any concerns before experiencing a loss event due to a breakdown in operational discipline, and before losing key drivers in the unrelenting quest for excellence in human performance. Any weaknesses revealed by the management review should be addressed in the fourth step of the PDCA cycle.

7.4.4.3 Self-assessment

Self-assessments are another tool that organizations can use to evaluate their level of operational discipline and identify areas for improvement. Self-assessments should be completed by workers who represent a cross-section of the site based on role and function, such as operations, management, technical, EH&S resources, operators, mechanics, and support staff. Self-assessments may be completed by individuals or by teams.

Electronic survey tools are particularly convenient for self-assessment surveys. They are inexpensive and allow regular sampling of worker opinions across the organization with instantaneous results. Analyzing the responses by job function, work group, site, business unit, seniority, etc., allows comparison of the responses (e.g., operators vs. supervisors) when interpreting the data. The survey results often provide a leading indication of opportunities for COO/OD improvement, but unexpected results should be verified with live interviews and/or reconciled with data from other COO/OD metrics before any corrective actions are undertaken.

An example from DuPont's self-assessment questionnaire is shown in the box that follows (Ref. 7.8). The entire questionnaire has seventy-five questions, and the answers are used to qualitatively score ten characteristics of operational discipline. Each operational discipline characteristic is assigned a section (e.g., Housekeeping), and one of the Critical-to-Quality elements is identified (such as "Personal value and demonstration of housekeeping standards by each individual"). The critical factors for each element are then scored on the following scale:

1. Not addressed in current state
2. Significant gaps or components missing
3. Partially in place with multiple opportunities for improvement
4. In place with minor opportunities for improvement
5. Fully in place with strong results

Personal value and demonstration of housekeeping standards by each individual

Gives status and recognition for good housekeeping. Reinforces link between good housekeeping and excellence in SHE results. **1 2 3 4 5**

Consider:

- Is everyone involved in housekeeping in their own areas or only a critical few?
- How does the site or area recognize excellence in housekeeping?
- Is there a periodic or annual focus on housekeeping?
- Is housekeeping evaluated and included in incident investigations and reports, where applicable?
- What percentage of time do individuals spend on personal housekeeping in their area?

Once the results are compiled, the site can focus its improvement efforts on elements with lower scores. Typically, improvement opportunities are also ranked so that the system improvements that will have the greatest impact on site operational discipline can be given top priority. Results and recommendations from the self-assessment are reported to site management for follow-up in the **Adjust** phase of the PDCA cycle. Results are primarily intended to help sites improve operational discipline locally; however, corporate programs could be developed for elements that score lower across multiple sites or regions.

7.5 ADJUST THE PLAN AND CONTINUOUSLY IMPROVE

The fourth step in the PDCA cycle is to analyze what is or is not working and make corrections or other necessary changes to the plan with respect to current goals. If the plan is meeting current goals, then the goal should be advanced to the next milestone in the quest for continuous improvement.

7.5.1 Evaluate Current Status and Gaps

Gap analysis is simply the process of comparing an organization's standards to workers' actual performance and identifying any gaps between the two. The following is the sequential gap analysis process:

1. Identify the relevant performance standards.
2. Have managers/supervisors observe (or review records that show) their employees' performance with respect to the standards. Determine whether the actual performance falls short of the desired performance. If so, there is a gap that must be addressed.
3. Determine the deficiency in knowledge, skill, or ability that caused the gap. There are several tools and techniques that can be used to (a) identify the source of the gap and (b) determine which specific skills and knowledge should be developed or improved in an effort to bring the employee's performance closer to the standards of the organization. Examples of these include the following:
    a. Training needs surveys – survey employees by conducting personal interviews or through written questionnaires.
    b. Customer feedback – gather informal and unsolicited customer complaints. Also use comment cards or formal interviews.
    c. Management observation – watch employees perform their job duties to determine gaps.
    d. Employee surveys – create surveys that ask employees how they think the organization is measuring up to the standards as a whole. The accompanying online material contains an example COO survey developed by Concord Associates. Surveys of organizational culture may also provide insight into underlying issues that may not be evident from surveys on COO/OD alone.

   e. Inspections – perform internal inspections in the same manner that government agencies examine organizations.
   f. Employee meetings – hold a roundtable or town hall meeting where employees informally discuss their concerns relative to areas for improvement, things they do well, and methods for improving performance.
   g. Audit results – create an operational audit checklist and compare the actual performance to the required standard.
   h. Incident investigations – evaluate the number of incidents where employee issues are the root cause.
4. Determine how to best remediate the deficiency. If the issue is a matter of knowledge or skill, then training, coaching, and/or mentoring may solve the problem. If the issue is a lack of ability or poor attitude, then resolution may require more fundamental changes in hiring, retention, and job-assignment policies.
5. Prioritize corrective actions. The prioritization should consider both the size of the gap between actual and desired performance as well as the importance of correcting the gap in achieving the organization's performance goals.

The culture and changes within an organization will dictate how often a gap analysis should be performed. If an organization experiences high turnover, if the gap between the actual and the desired performance required significant retraining or other remediation, or if there has been major organizational changes, then a gap analysis should be performed more frequently. On the other hand, if an organization is relatively stable, then a gap analysis that is performed annually or semiannually may be adequate to monitor employee effectiveness.

7.5.2 Common Implementation Problems

It is inevitable that there will be gaps during the early years of COO/OD implementation, and the gaps will likely vary for different business units, sites, and work groups. There will always be gaps when management advances its expectations to the next milestone toward the organization's ultimate goal. The following are frequently observed gaps in COO/OD implementation, some of which will be hard to spot unless management is specifically looking for them:

- Performance standards are created and applied just prior to the performance appraisal and are written to match current performance. This will give a false impression of excellent compliance. Performance standards should be applied at the beginning of the appraisal period so that current performance will fairly reflect change during the period.
- Supervisors are micromanaging employees beyond the requirements of the performance standards. Employees should be empowered to use different methods to get the same results within the requirements of the standards.

- Performance standards are never updated or revised, so they are either easy to exceed (older, less demanding requirements) or not appropriate (different requirements). Performance standards should be flexible and evolve with the job requirements and the organization's goals. The standards should be periodically assessed to ensure alignment with the organization's current goals, and they should be updated as job requirements change.
- The performance standards are not being implemented due to disagreement between the supervisor and the employee. It is always best if the front-line worker agrees with the performance standards, but the manager or supervisor has the final say as to what standards are suitable. The worker may be reluctant to change due to peer pressure, which the supervisor must address as a broader issue.
- The current standards are not specific or measurable. Thus, when performance is evaluated, the judgments are necessarily subjective and will likely be ambiguous. The measurement gap must be corrected before a performance gap can be fairly determined.
- The current plan does not clearly specify which metrics will be collected to ensure that the performance standards are achieved. Thus, the metrics used to reveal gaps may not accurately reflect current performance.
- The current standards are not realistic, given the organization's current situation. Specifying performance that is clearly beyond the grasp of an employee is unreasonable and frustrating for all involved. In this instance, gaps are created by the standard writers, not the front-line workers.
- The supervisor fails to provide adequate supervision. The supervisor must provide guidance, training opportunities, leadership, motivation, and the proper role model. When this is not the case, focusing on the front-line worker will not correct the performance gap.
- Activities are not appropriately planned. For example, the operational tempo and/or schedule puts individuals at unacceptable risk (e.g., due to insufficient rest or staffing), and performance is adversely affected. Such situations, which might unavoidably arise during emergency situations, are unacceptable during normal operations.
- The supervisor fails to correct known deficiencies among individuals, equipment, training, or other safety-related areas. For example, the failure to consistently correct or discipline inappropriate behavior certainly fosters an unsafe atmosphere, but it may not be evident in the performance metrics if no specific rules or regulations are broken. This problem is compounded if the supervisor shows bias or favoritism and only corrects the behavior of "problem" workers.
- The supervisor willfully disregards existing rules and regulations when managing assets. For instance, permitting an individual to operate a forklift without current training or qualifications sets the stage for a tragic accident, but current performance may appear acceptable.

7.5.3 A Maturity Model for COO/OD (Current Status and How to Proceed)

In addition to gaps identified by comparing the COO/OD system to the organization's current standards, there may be gaps between the organization's standards and industry best practices. A COO/OD maturity model, such as the one summarized in Table 7.2, can serve as an independent basis for comparison. An organization contemplating a new COO/OD system would likely rate itself as Stage 1 or Stage 2 in the maturity model, while an organization looking to improve an existing COO/OD system might rate itself as Stage 3 or Stage 4.

To achieve the full benefits of a COO/OD system, the organization needs to strive for Stage 5 implementation throughout. The maturity model offers milestones for improvement, even if the organization has no gaps in its current implementation. Management can target the next level of maturity as its goal and adjust its plans for the next PDCA cycle accordingly.

7.5.4 Prioritizing Improvement Opportunities

If the results of the gap analysis show any significant variance from the desired progress in achieving the organization's goals, management should develop a corrective action plan. Organizations with mature COO/OD systems should have relatively few gaps, and those that are discovered can be addressed directly. But organizations with less mature systems may find so many gaps that they are overwhelmed with opportunities for improvement. In those cases, management must set realistic goals and priorities for achieving them.

The most successful strategies for implementing COO/OD build on successes and lessons learned from experience. So management should first identify those things that are being done well within the COO/OD system and communicate them throughout the organization. From those successes, management can extract a toolbox of best practices that can be shared across the organization and used to close any gaps that were found.

The next step is to sort the identified gaps into major and minor deficiencies. The minor deficiencies should be addressed directly. Corrective action may require no more than coaching or retraining a few individual(s) or getting a work group to adopt the same practices that have proven successful elsewhere in the organization.

Correcting a major deficiency, or pushing the organization to the next level in the maturity model, will require more careful planning. As when COO/OD was originally introduced, the best strategy is to start with smaller efforts that have a high probability of success. Prioritize the effort by applying the concepts of quality pioneer Dr. Joseph Juran and focus on the "vital few" changes that will have a large impact on results rather than the "trivial many" that could consume most of your energy for minor gains. (Others refer to the Pareto Principle, which says that 80% of the gain will be realized from 20% of potential investments.) Unfortunately, the process safety benefits of a change may not be obvious, so management must make an informed assessment based on anticipated risk reduction.

TABLE 7.2. Stages in the Evolution of a COO/OD System

Stage	Leadership and Commitment	Standards and Procedures	Metrics
5 Continuously Improving	Leadership drives the process for COO/OD to achieve the organization's safety goals Leadership gives visible, unconditional support with constant communication Employees own COO/OD processes and coach peers Employees believe that the organization is committed to safety via COO/OD	Work teams are empowered to upgrade standards and procedures Work teams network and share best practices across the organization	Leading indicators are primarily used New metrics are selected to reveal improvement opportunities Lagging indicator trends confirm ongoing improvement
4 Management System in Place	Leadership includes safety in its strategic plans on an equal footing with financial performance Leadership has clearly stated goals and constancy of purpose Supervisors and managers are championing COO/OD processes, empowering employees, and leading the development of team goals Employees are assuming leadership roles in COO/OD Employees believe that safety is a core value	A process is in place to ensure that standards and procedures are current Employees train others in COO/OD	Leading indicators are primarily used Lagging indicators confirm COO/OD success

TABLE 7.2. Stages in the Evolution of a COO/OD System

Stage	Leadership and Commitment	Standards and Procedures	Metrics
3 Implementing	Leadership sets goals for improving safety performance Leadership is actively communicating safety expectations and improvement goals to front-line employees Supervisors and managers execute elements of the COO/OD system on a planned timeline Supervisors are assuming COO/OD leadership roles Commitment to safety is a condition of employment Employee safety concerns are identified and addressed	A defined work process is in place for the development, maintenance, and review of standards and procedures Employees are involved in developing COO/OD standards and procedures	Leading indicators are defined and being collected, and they are being used to drive COO/OD activities Sites/facilities are measuring their progress in COO/OD implementation
2 Program Development	Leadership proactively attempts to improve safety performance by closing gaps in the safety management system Leadership discusses safety in meetings with managers, supervisors, and front-line employees Accountability for safety is discussed primarily during performance appraisals or after incidents occur	Standards and procedures are consistently enforced across each site An undocumented process exists for developing and reviewing standards and procedures Employee input on standards and procedures is solicited	Leading indicators are defined and being collected, but they are not being used to drive COO/OD activities Sites/facilities use audits to measure COO/OD implementation

TABLE 7.2. Stages in the Evolution of a COO/OD System

Stage	Leadership and Commitment	Standards and Procedures	Metrics
1 Awareness Building	Employees believe that business performance is valued above safety		
	Leadership has no specific goals for future safety performance	Supervisors inconsistently enforce standards and procedures – the emphasis is on results	Lagging indicators (e.g., injury/illness rates) are used
	Leadership addresses COO/OD issues in reaction to incidents	Employees rely on supervisors for standards and procedures	

Introduce the revised plan on a pilot scale using work groups that are already enthusiastic about the benefits of COO/OD. Then close the PDCA loop by communicating their successes to others and adjusting their approach (as necessary) to serve as a model for other work groups or locations to follow.

7.6 APPLICATION TO DIFFERENT ROLES

COO/OD applies to all departments and all levels of the organization: management, operations, technical support, administrative support, and hourly personnel. Inherently, the phrase conduct of *operations* implies that the operations department will be heavily involved. But the same principles can be applied by other work groups to attain the goals of the organization.

The research and development (R&D) group must invent or adapt new products and processes that will satisfy customers' needs. The standards of performance applicable to this group are mostly focused on their work methods with the belief that disciplined application of the methods will produce the desired results. Research, by its very nature, involves investigating the unknown, but proper application of COO/OD principles reduces the safety risk to workers and assets, as well as the risk of experimental failure. For example, R&D personnel must perform energy release calculations prior to performing experiments, safety equipment must be operational whenever experiments are conducted, and experimental protocols must be rigorously followed. Thorough R&D will define safe operating limits that will be incorporated into the performance standards for the operations department.

The engineering department also has a key role to play in implementing a COO/OD system. When designing, installing, or modifying equipment, engineering should facilitate operations' compliance with COO/OD requirements. Even the best COO/OD system cannot offset fundamental flaws in facility or equipment design; therefore, engineering should seek operations' input on ways to optimize the design with respect to COO/OD. For example, equipment can designed and arranged to facilitate easy cleaning and routine maintenance. Proper labels can be affixed to the equipment, and clear operating and maintenance procedures can be delivered with the equipment. Whether for a small modification or a major capital project, the engineering activities should themselves be conducted in accordance with COO/OD principles so that projects are completed on time and within budget, and operate as expected when commissioned. For example, documents and revisions must be controlled, safety analyses must be performed on finalized designs, the latest versions of codes and standards must be used, and the facility and/or equipment must be built to specification.

Outside of operations, maintenance is the department most directly affected by COO/OD. Operations depends on preventive and corrective maintenance activities being performed in a timely and accurate fashion. Equipment must be prepared for maintenance and returned to service afterwards, so accurate communications between the two departments are essential. The nuclear industry developed a special version of COO/OD for maintenance activities called STAR (Stop, Think,

Act, and Review) because of the number of unplanned unit trips caused by maintenance errors.

Engaging all departments in COO/OD will enhance performance in process safety, personal health/safety, environmental responsibility, quality, productivity, and profitability.

Any organization can improve profitability by simply implementing and performing its operations and other job functions with discipline. The benefits accrue from:

- Better alignment among operations, engineering, maintenance, R&D, and other business functions
- Greater use and greater efficiency of employees' skills, knowledge, and ability
- Higher quality products without increased cost
- Increased capacity without higher capital spending
- More efficient use of all resources: human, capital, assets, and technology

7.7 SUMMARY

Implementing and maintaining an effective COO/OD system is a proven way to improve the performance of any organization. Its importance has grown as systems have become more complex and the consequences of failure more severe. As discussed in Chapter 2, industries as diverse as manufacturing, aviation, healthcare, and defense have met aggressive goals and improved their performance by orders of magnitude by systematically implementing COO/OD systems. Individual companies in the process industries have also demonstrated outstanding performance with COO/OD systems, and the time has come for these methods to be more generally adopted.

As discussed in Chapter 3, the leadership team must seize the initiative and commit itself to the implementation of COO/OD. As discussed in Chapter 5, there are many facets to a truly comprehensive COO system, and it will ultimately affect every work activity. Similarly, as discussed in Chapter 6, the commitment to operational discipline will improve human performance in every work activity. However, the path to an effective COO/OD system is not simple, and it cannot be implemented quickly – but it is achievable. The same PDCA cycle used to implement other organizational changes can also be used to implement COO/OD. The NUMMI experience, and others described in this book, should inspire those who want to improve process safety with COO/OD – same people, same plant, different management, astonishing results.

7.8 REFERENCES

7.1 Shook, John, "How to Change a Culture: Lessons from NUMMI," *MIT Sloan Management Review*, Massachusetts Institute of Technology, Cambridge, Massachusetts, Vol. 51, No. 2, Winter 2010, pp. 63-68.

7.2 Klein, James A., "Operational Discipline in the Workplace," *Process Safety Progress*, American Institute of Chemical Engineers, New York, New York, Vol. 24, Issue 4, December 2005, pp. 228-235.

7.3 Sharpe, Virginia A., "Promoting Patient Safety: An Ethical Basis for Policy Deliberation," *Hastings Center Report Special Supplement*, The Hastings Center, Garrison, New York, Vol. 33, No. 5, September-October 2003, pp. S1-S20.

7.4 Center for Chemical Process Safety of the American Institute of Chemical Engineers, *Guidelines for Process Safety Metrics*, John Wiley & Sons, Inc., Hoboken, New Jersey, 2009.

7.5 U.K. Health and Safety Executive, *Developing Process Safety Indicators: A Step-by-Step Guide for Chemical and Major Hazard Industries*, HSE Books, London, England, 2006.

7.6 U.K. Health and Safety Executive, *A Guide to Measuring Health & Safety Performance,* London, England, December 2001.

7.7 Klein, James A., and Bruce K. Vaughen, "A Revised Program for Operational Discipline," *Process Safety Progress*, American Institute of Chemical Engineers, New York, New York, Vol. 27, Issue 1, March 2008, pp. 58-65.

7.8 Klein, James A., and B. K. Vaughen, "Evaluating and Improving Operational Discipline," Process Plant Safety Symposium, Houston, Texas, April 22-26, 2007.

7.9 ADDITIONAL READING

- Dekker, Sidney, *Just Culture: Balancing Safety and Accountability*, Ashgate Publishing Company, Burlington, Vermont, 2007.

INDEX